Soviet Oil and Gas to 1990
and
The Market for LPG in the 1980s

Soviet Oil and Gas to 1990
by David Wilson

and

The Market for LPG in the 1980s
by Geoffrey Drayton

Abt Books
Cambridge, Massachusetts
EIU Special Series 2

Library of Congress Cataloging in Publication Data

Wilson, David.
  Soviet oil and gas to 1990.

  (Economist Intelligence Unit special series)
  "Economist Intelligence Unit Ltd."
  1. Petroleum industry and trade—Soviet Union—Forecasting. 2. Gas industry—Soviet Union—Forecasting. I. Drayton, Geoffrey. II. Economist Intelligence Unit Limited. III. Title. IV. Series.
  HD9575.S652W54   1982   333.8'2311'0947   82–13850
  ISBN 0–89011–581–8

Originally published by The Economist Intelligence
Unit as Special Report Nos. 90 and 96.

Printed in the United States of America

The Authors

David Wilson holds a Ph. D. in Russian Studies and a Master of Social Science degree in national economic planning from the University of Birmingham. He is a Lecturer in Soviet geography at the School of Geography, Leeds University. He is a regular contributor to EIU's <u>Quarterly Energy Review Of The USSR</u> + <u>Eastern Europe</u>, and has written a number of reports on the Soviet energy industry on a private consultancy basis.

Geoffrey Drayton is Manager of Energy Projects in the EIU's Economic Studies Department and as such has ultimate responsibility for all research in the energy field. He is also a regular contributor to EIU's <u>Quarterly Energy Review</u>.

Contents

List of Tables

Soviet Oil and Gas
to 1990

Introduction

Ever since the USSR overtook the USA in 1974 to become the world's largest oil producer, speculation has mounted over its ability to remain self-sufficient in oil, provide the major share of its East European allies' requirements, and obtain a large part of its hard currency needs through sales to the West.

In April 1977, the CIA published the first of its controversial reports on Soviet oil. It predicted a peak in output of 11 to 12 mn barrels a day (550 to 600 mn tons/yr) by the early 1980's, followed by a rapid decline to between 8 and 10 mn b/d (400 to 500 mn tons/yr) in 1985. This would necessitate imports on the order of 3.5 to 4.5 mn b/d by the Soviet Union and Eastern Europe in 1985, and would place a considerable strain on world oil supplies. While this view has now been abandoned by the CIA, the debate over future Soviet capabilities and intentions has not been resolved, and a wide variety of viewpoints remain over the nature and scale of Soviet involvement in the world energy markets.

At the present time, the USSR is an important net exporter of oil and gas. Sales of oil and oil products have remained at about 160 mn tons/yr for the last few years and have held up well during the Western recession, although a slight decline may be anticipated in the near future. Exports of natural gas should rise substantially from the 57.3 bn cu m of 1980, particularly during the second half of the 1980's. The Russians have endeavoured to peg their oil exports to Eastern Europe at about 80 mn tons/yr while delivering increasing volumes of gas and electricity. This has enabled them to maintain oil sales to OECD countries at about 55-60 mn tons/yr while exports of natural gas to these markets is also growing. A further 10-15 mn tons/yr of oil is sold to developing countries, principally Cuba, Vietnam, India and Brazil.

When hydrocarbon exports are considered in value terms, it can be seen that the USSR, currently the world's largest exporter of gas and second largest exporter of oil, has benefited greatly from the oil price increases of 1979 and 1980. The value of its oil and gas exports rose by 88.9% over the period 1978-1980 and sales to the "hard-currency West" rose by 101%. This enabled it to achieve a trade surplus of $5.3 bn with European OECD countries in 1979, followed by a surplus of $7.5 bn in 1980, and these have been sufficient to pay for large grain imports from the USA. With the OECD as a whole, the USSR has run a surplus of $0.9 bn in 1979 and $3.2 bn in 1980.

This change in the USSR's trading fortunes is entirely due to its exports of oil and gas. In 1980 they accounted for 63 per cent of the value of its exports to OECD countries (up from 62 per cent in 1979) and it can be said that they play a major role in the development of the whole Soviet economy.

If the USSR were no longer able to export oil and gas, the negative effect on its economy would be substantial. There is little doubt that it would be unable to pay for significant imports of fuel, and certainly not on the scale envisaged by the CIA. It can be said with equal conviction that the Soviet leadership would never allow the country to become dependent on external supplies of an important commodity like oil, and would make every conceivable effort to avoid such a situation.

Yet there is no sign that the Russians appear unduly worried. While investment and capital construction programmes have been stepped up in the most important fuel-producing region of Western Siberia so as to ensure plan fulfilment, there is no evidence of the switching of resources from other sectors which would be inevitable in the event of an impending oil crisis. Enormous prestige projects of dubious short term value, such as the BAM railway, are proceeding at planned rates, soaking up vast material and labour resources that could be sent to Western Siberia if there were real concern about future oil supplies.

It has been suggested by Western journalists that the bureaucratic nature of the Soviet government is such that the Russians could find themselves in the midst of an oil crisis before they realised it. "Astonishing to relate, the Soviet Union does not even have a coordinating Energy Ministry."[1] In fact, policy for the fuel-producing sector is elaborated by a special committee of the Council of Ministers headed by Gosplan chairman Nikolai Baibakov, a former minister of the oil industry. The committee consists of the ministers of all the 14 ministries involved in various aspects of the fuel industry as well as political leaders from fuel-producing regions and representatives from scientific institutions and re-search organisations. It is a shadowy organisation whose proceedings are top secret, but it is likely that the inter-ministry rivalries and struggles for funds are reflected in the meetings of the committee.

It has also been argued by Western journalists that the advanced age of Soviet politburo members makes radical shifts in policy impossible to effect. In fact, it is this Council of Ministers' committee which formulates policy; President Brezhnev's role was latterly confined to holding the various interest groups together through his unchallengeable authority. The policy debates may now become more lively, but it is unlikely that ministers will lose sight of the importance of the job in hand.

Far from being impervious to an impending doom foreseen by some in the West, the Soviet government monitors the progress of the sectors producing, transporting and processing oil and gas more closely than any other sector of the economy, apart from the military.

The oil and gas sectors are increasingly interlinked

Oil and gas are being considered together in this report because of their increasing inter-relationship. Since the formation in 1973 of the Ministry of Construction of Oil and Gas Enterprises (Minneftegazstroi), cobbled together from the relevant parts of the Ministry of Oil and Ministry of Gas, Soviet planners have been able to alter the capital construction balance between oil and gas much more freely. This process is facilitated by the fact that one ministry, the Ministry of Chemical Engineering (Minkhimmash), is responsible for the manufacture of equipment for the production of both oil and gas.

Consequently, pipeline builders employed by Minneftegazstroi can find themselves working on either oil or gas pipelines. Similarly, the workers of the Siberian Modular Assembly Organisation (Sibkomplektmontazh) can be installing a gas treatment plant one month and an oilfield pumping station the next. A graphic illustration of the multi-purpose nature of equipment is the use of the Suleiman Vezirov pipelaying ship (usually laying oil pipelines in the Caspian Sea) to lay a gas pipeline across the Kuibyshev Reservoir.

From the fuel consumption aspect, the substitutability of oil and gas is becoming more apparent. Most new fuel-burning power stations introduced during the last 20

---

1 Adrian Berry, Daily Telegraph, March 19, 1980

years have been designed to operate on two or more fuels. One reason for this is the seasonality of demand for gas, with many gas-burning facilities such as municipal boilers and combined heat-and-power plants operating at only half capacity in summer. This has meant that gas production must fall markedly during the summer months, in spite of the construction of a number of underground gas storage reservoirs. But as transportation accounts for by far the greatest share in the delivered cost of gas, it is preferable to keep gas pipelines operating at optimal capacity. Consequently, oil-fired power stations and boilers are designed so as to serve as "buffer consumers" of natural gas in summer.

One effect of this has been for the seasonality of demand for fuel oil to become more accentuated. This can be seen from the big rise in daily crude output rates during the last three months of most years, which enables stocks of fuel oil for power stations to be built up for winter. It can also be seen from the sharp fall in oil exports during the first quarter of every year.

The increasing use of natural gas for electricity and heat generation has created conditions in which the share of fuel oil in the total output of oil products can be reduced. It currently stands at about 50 per cent compared with 35 to 40 per cent in Western Europe and 13 per cent in the USA, where a half of all crude is refined down to petrol. The extent of the possible substitution of crude oil by natural gas is thus determined by the speed with which the USSR can bring on-stream sufficient secondary refining capacity. This will enable a greater volume of light and middle products to be obtained from a given volume of crude, thereby reducing the output of heavy fuel oil.

The likely extent of substitution of natural gas for oil is one of the major imponderables of Soviet fuel forecasting. In predicting Soviet import requirements of 3.5 to 4.5 mn b/d by 1985, the CIA did not believe that substitution could take place to any significant extent over a period of eight years. I too underestimated the scale of Soviet intentions in my report "Soviet Oil and Gas to 1990" published by the Economist Intelligence Unit in December 1980, but written before the 11th Five Year Plan (FYP) targets for 1985 were announced. My calculations that the USSR was planning to develop capacity for the output of 700 mn tons of oil and 600 bn cu m of gas in 1985 (the actual plan targets have been set at 630 mn tons of oil and 630 bn cu m of gas) implied a modest growth in secondary refining capacity in keeping with the Soviet record to date. However, Soviet planners have evidently budgeted for a substantial growth, and if their plan for the installation of secondary refining equipment is fulfilled, then this should allow domestic demand for light and middle products to grow at an annual rate of 4 ot 5 per cent from an annual increase of less than 1 per cent in the production of crude oil, with a slightly declining level of exports.

It has been suggested in the West that, rather than accelerate the substitution process in their domestic economy, the Russians will phase out their oil sales to the West in favour of gas during the next few years. It should be remembered, however, that transcontinental pipelines are the only practicable method of delivering gas to Western Europe, and the Russians cannot expect a big rise in gas exports until 1985 when the throughput of the Yamal pipeline begins to build up. Therefore it is possible to assume that most of the 45% increase planned for gas output over 1981-1985 will be used domestically with a simultaneous fall in fuel oil consumption. Moreover, the need to pay for grain imports, service debts to Western banks, and bail out Poland will require the Soviets to push oil sales (products rather than crude) to the West as hard as they can, and they have therefore embarked on a crash programme of gas for oil substitution in their domestic economy on a quite extraordinary scale.

So the future of the oil production sector is inevitably bound up with the growth of the secondary refining sector and the capacity of natural gas to make up for the declining output of fuel oil. The ability of the USSR to meet the demand for oil by itself and its allies, and to maintain its exports of oil to the West, is thus directly linked with the future of its natural gas industry. For this reason, these two fuels are being considered together in this book.

CHAPTER 1

The Soviet Fuel Balance

In 1950, coal accounted for 66 per cent of Soviet fuel production, with other low-grade fuels such as oil shale, peat and firewood providing a further 14 per cent. During the 1950s, the share of oil and gas nearly doubled, from 20 per cent to 38 per cent, and then rose further to 60 per cent during the 1960s. By 1975, oil accounted for 44.7 per cent, gas 21.8 per cent, coal 30 per cent and other fuels 3.4 per cent of total output.

During the Tenth five year plan (FYP) period of 1976 to 1980, the share of oil in the fuel balance stabilised at 45 per cent as expected by the plan targets for 1980. The share of gas was higher than its plan target of 25.6 per cent while that of coal was correspondingly lower. This is because natural gas was the only fuel to meet its Tenth FYP target of 435 bn cu m. Oil fell short of its 640 mn ton target by 5.8 per cent, and coal by 11 per cent (target 805 mn tons, actual output 716 mn tons).

The Tenth FYP called for 2,035 mn tons of standard fuel[1] to be produced in 1980. In the event, the Soviets produced 1,906 mn tons, amounting to an increase of 21.3 per cent over the 1,571 mn tons of 1975 compared with a growth of 28.6% during the previous five year period 1971-1975.

The 26th Congress of the Communist Party, held in February 1981, called for the output of the three most important fuels (oil, gas and coal), to grow by 14.1-19.2 per cent during 1981-1985. The Party's November 1981 Plenum confirmed a target of 16.6%, equivalent to 3.15 per cent a year, although the actual increase obtained in 1981 amounted to only 1.88 per cent. This means that an average growth of 3.43 per cent a year must be obtained during 1982-1985 for the target to be met.

The targets approved by the November Plenum include 630 mn tons of oil (the Congress set a target range of 620 to 645 mn tons), 630 bn cu m of gas (600 to 640) and 775 mn tons of coal (770 to 800 mn tons).

---

1 Measured at 7 mn keals/ton. 1 ton of oil = 1.42 tsf, 1,000 cu m of gas = 1.19 tsf.

Table 1

Soviet Fuel Balance 1960 - 1985 (mn tons of standard fuel)

Volume of output

	1960	1970	1975	1980	1981[1]	1985 Plan[2]
Oil	211	502	702	863	871	901
Natural gas	54	234	343	516	551	747
Coal	373	433	472	484	476	524
Others	55	53	54	43	40[3]	40[3]
Total	693	1,222	1,571	1,906	1,938	2,212

Share of Total

	1960	1970	1975	1980	1981	1985 Plan
Oil	30.4	41.1	44.7	45.3	44.9	40.7
Natural gas	7.8	19.1	21.8	27.1	28.4	37.8
Coal	53.8	35.4	30.0	25.4	24.6	23.7
Others	7.9	4.3	3.4	2.2	2.1[3]	1.8[3]
Total	100	100	100	100	100	100

Sources: Narodnoe khozyaistvo 1980. 1. Pravda 24/1/82 2. Izvestia 3. Assumed

Natural gas has accounted for 50 per cent and oil for 45 per cent of the total increase in fuel production over the period 1976-1980. The contribution of coal and other fuels has been negligible. This situation is likely to persist in the future with gas becoming even more important, but the contribution of coal to the increase in fuel output over the 1980s should rise slightly as the development of the Kansk-Achinsk and Ekibastuz basins intensifies.

As may be deduced from Table 2, the production of oil and gas has continued to grow over the period 1975-1981, although the growth rates for oil production have fallen in both percentage and absolute terms.

Table 2

Growth of output of oil and gas 1975 - 1981

	Oil			Gas		
	Output (mn tons)	Growth over previous year. %	mn tons	Output (bn cu m)		
1975	490.8			289.3		
1976	519.7	5.9	28.9	321.0	11.0	31.7
1977	545.8	5.0	26.1	346.0	7.8	25.0
1978	571.5	4.7	25.7	372.2	7.6	26.2
1979	585.6	2.5	14.1	406.0	9.2	34.4
1980	603.2	3.0	17.6	435.2	7.0	28.6
1981[1]	609	1.0	5.8	465	6.8	29.8
1982 Plan	614[2]	0.8	5.0	492[3]	5.8	27.0

Sources: Narodnoe khozyaistvo   1. Pravda 24/1/82   2. Tass 27/1/82 3. PTO.

The reason for the low and declining growth rates for oil can be viewed from both a consumption and a production angle. The current policy of substitution of natural gas for fuel oil, especially in summer, has already been mentioned, and the slow growth in oil output can be seen as a corollary of the rapid growth in gas production. So, despite an increase in oil output of only 1 per cent in 1980, the production of light products has been sufficient to satisfy demand while oil exports have held up.

From the production side, the problem is that Western Siberia is now having to provide the whole increment in national output as well as cover the decline in the rest of the country. While this factor was foreseen by the planners, the decline in production in and around the Caspian Sea has been greater than they expected, as will be explained below.

Table 3 shows how the output of oil from regions other than Western Siberia has fallen from a peak of 340 to 350 mn tons in the mid 1970's to 274 mn tons in 1981.

Table 3

The Growing Importance of Western Siberian Oil Production (mn tons)

	USSR	Western Siberia	Rest of the USSR	Western Siberia as a share of USSR, %
1965	242.9	1.0	241.9	0.4
1970	353.0	31.4	321.6	9.8
1975	490.0	148.0	342.0	30.2
1979	585.6	284.4	301.2	38.6
1980	603.2	313.5	289.7	52.0
1981	609	335	274	55.0
1982 Plan	614	355	259	57.8

While output targets for the two most important producing regions, Western Siberia and the Urals-Volga region, are being regularly fulfilled, funds and resources are diverted from the lesser producing regions where production is now falling rapidly.

The gas industry is one of the select few which achieved their 10th FYP targets. Annual plans have been consistently over-fulfilled, and again it is Western Siberia which is accounting for most of the annual increase in output. The dominance of Western Siberia is not yet as great as for oil - only 40 per cent in 1981 - but it is planned to grow to nearly 60% by 1985.

Fuel and economic development

During the period 1950-1980, Soviet industrial production grew by 1122% at comparable prices, while the output of the three major fuels rose by 598%. Thus each 1% increase in fuel output accompanied a 1.88% increase in industrial production. During the 5 years 1976-1980, however, each 1% rise in fuel output was accompanied by a rise of only 1.05% in industrial production. This is explained partly by the fact that fuel exports rose rapidly to pay for imports of agricultural produce which made no contribution to the growth in industrial production. But it also reflects a slowdown in the rate at which fuel utilization by industry is improving, and this explains the emphasis on fuel saving during the current plan period. Gosplan chief Baibakov told the November Plenum that 200 mn tsf must be saved over 1981-1985 compared with 125 mn tons saved during 1976-1980. If these savings are achieved, then the 11th FYP target of a rise of 1.67% in industrial production per 1% increase in fuel output should prove feasible, and fuel exports should be maintained at current levels. The fuel saving programme seems to have got off to a good start; in 1981, industrial production rose 3.4% with a rise in fuel output of 1.9%.

The atomic power programme is expected to play a major role in reducing the dependency of industrial growth on fuel production. It is planned to produce 1,555 bn kwh of electricity in 1985 compared with 1,295 bn in 1980, and output by atomic and hydro plants will rise to 450 bn kwh compared with 250 bn in 1980, thereby accounting for 77% of the increase. But the atomic programme has already run into difficulties. The 11th FYP has called for the commissioning of 23 to 25 GW of atomic capacity, practically all of it at grassroots stations with all the attendant infrastructure problems. During 1981, only 2.88 GW were installed, with the construction of vitally important units at Smolensk, Nikolaev and Kursk lagging badly. Unless there is a dramatic improvement in the atomic construction sector, then the Russians will be forced to rely even more heavily on their Siberian oil and gas fields to make up the shortfall in energy.

CHAPTER 2

The Location of Production of Oil and Gas

<u>WESTERN SIBERIA</u>

OIL

It can be safely assumed that the future of Soviet oil production during the next
ten years hinges on the performance of the Western Siberian fields located in
Tyumen and Tomsk oblasts. Holding perhaps 75 per cent of the country's proved
and probable reserves and containing some of its largest deposits, Western Siberia
is expected to account for 63 per cent of Soviet oil in 1985 and up to 75 per cent
in 1990.

Oil was discovered in the region in 1960, and production began at the small Shaim
field in 1964. By 1970, when 31 mn tons were produced, the massive Samotlor
deposit had come on-stream, and output then rose rapidly to 148 mn tons in 1975.
The Tenth FYP target was set at 315 mn tons for Western Siberia, and this was
almost fulfilled, with Tyumen oblast producing 302.2 mn tons and Tomsk oblast
10.8 mn tons.

The development of the Western Siberian fields has been bedevilled by uncertainty
about their potential. In 1969, for example, a conference on Siberian resources in
the city of Akademgorodok was told that Siberian oil would never account for more
than a third of total Soviet output, and in 1970, when the Ninth FYP target for
1975 was set at 125 mn tons, it was asserted that output in 1980 would range
between 230 and 260 mn tons. As it became apparent that the 1975 target was going
to be substantially overfulfilled, predictions for 1980 began to rise: to 260-300
mn tons in 1973 and 300-330 mn in 1974, and the Tenth FYP target was set at 315 mn
tons.

In 1974, a target of 400 mn tons for 1985 was made, and the 11th FYP has called for
the production of 395 mn tons, 26.6% more than in 1980. In 1981, the annual plan
target of 335 mn tons was almost fulfilled, and a target of 355 mn tons, including
343 mn tons by Tyumen oblast and 12 mn tons by Tomsk has been drawn up for 1982.
Looking ahead to 1990, the head of the Tyumen oblast committee of the Communist
Party Bogomyakov and the head of the Tyumen oblasts exploration organisation
Tyumengeologiya Farman Salmanov have both claimed that the oblast can provide 500
mn tons of oil, and perhaps even more depending on the speed with which some of the
more remote deposits can be developed.

The production of oil in Western Siberia is administered by the Glavtyumenneftegaz
Association in Tyumen oblast and the Tomskneft Association in Tomsk oblast.
Glavtyumenneftegaz consists of five associations: Nizhnevartovskneftegas, centred
on Nizhnevartovsk, Surgutneftegaz centred on Surgut, Yuganskneftegaz based in
Nefteyugansk, Noyabrskneftegaz located in the new town of Noyabrskii on the Surgut -
Urengoi railway, and Shaimneft which operated from Urai.

Table 4

Production of Oil in Western Siberia, mn tons

Tyumen oblast	28.0	143.0	274.4	302.2	323.5	380
Nizhnevartovskneftegaz	10.4^1	100.8^1	186^2	201.5^6	207.4^9	212
Surgutneftegaz	13.5^1	36.8	42^3	49^7	61.2	70
Yuganskneftegaz			39.6^4	45.5	50.7	50
Noyabrskneftegaz					4.2^{10}	36^{10}
Shaimneft	4.1^1	5.4^1	5.8^5	6	6	12^{11}
Tomsk oblast	3.4	5	10	10.5	10.5	15
Total	31.4	148	284	312.7	334.0	395

Sources 1: Muravlenko & Kremneva "Sibirskaya neft", Moscow 1977 p. 186.
2: Khanti-Mansiisk radio 28/2/80 3: Moscow radio 2/3/80
4: Khanti-Mansiisk radio 11/1/80 5: Estimate from Moscow radio 30/11/79
6: Estimate from Moscow radio 16/11/80 7: Prediction by Izvestia
15/10/80 8: Neftyanoe khozyaistvo No. 5, 1981 9: Est from Moscow
radio 22/9/81 10: Forecast by Izvestia 18/9/81. Previous data for
Noyabrskneftegaz is included under Surgutneftegaz 11: Izvestia 27/
10/81

Tyumen Oblast

The Nizhnevartovskneftegaz Association includes five oil-producing directorates (NGDU's) as well as three drilling directorates (UBR's) and others involved in transport, construction and services such as well repairs. The oil-producing directorates are: Nizhnevartovsk Lenin (which produced 85 mn tons in 1981) and Belo-ozersk (73 mn tons) which are jointly responsible for exploiting Samotlor: Megionneft (26 mn tons) which works Megion, Potochnoe, Urevsk, Pokursk, Pokacheva and Lokosovsk deposits; Varyeganneft (23 mn tons) which works the Varyegansk group of deposits; and Zapolyareneft which has not produced any oil yet, but which has been set up to organise the exploitation of the Russkoe in the far north of the oblast.

The Surgutneftegaz Association includes two production directorates: Surgutneft which works the giant Zapadno-Surgut deposit together with a cluster of small adjacent deposits, and Fedorovskneft which was split off from Surgutneft in 1977 to concentrate on Siberia's second largest producer of Fedorovsk.

Noyabrskneftegaz was created in July 1981 out of the Knolmogorneft directorate of Surgutneftegaz to administer the Kholmogorsk and Karamovsk deposits. During the current FYP period, the Muravlenkovsk and Sutorminsk deposits will also be brought on-stream, and the new Association has been set the task of producing 82 mn tons over 1981-1985.

The Yuganskneftegaz Association includes five production directorates: Yuganskneft, Pravdinskneft, Mamontovneft (which works Siberia's third largest deposit of Mamontovo and produced 26.3 mn tons in 1981), Malaya Obneft and Salymneft. The latter, in conjunction with Salym UBR, is responsible for working the Salym deposit.

The Shaimneft Association has only one production directorate, Uraineft.

Of the 43 deposits being worked at the beginning of 1982, only seven produce appreciable amounts of oil, and these are dominated by the supergiant Samotlor. It was discovered in 1975, and was quickly recognised to be the Soviet Union's largest oilfield with recoverable reserves initially set at a level of 2,000 mn tons. Development began in 1969 when 1.4 mn tons were produced, and output quickly rose to 86.2 mn tons in 1975 and 110 mn tons in 1976. The Tenth FYP expected Samotlor to peak at about 130 mn tons in 1977, but output continued to rise to 155 mn tons in 1980, and was maintained at this level in 1981.

How long this peak level of output can be sustained depends on the extent to which appraisal drilling reveals new pay-zones. Originally, five significant payzones were distinguished, and development began with the most productive BV_8 in the central part of the deposit. During the Tenth FYP period, AV_1 and AV_{2-3} were introduced. It was not intended to work BV_{10} until 1983 because of its geological complexity, but appraisal drilling revealed a highly productive section in its south-east corner, and it is now being developed. Western analysts have pointed out that only some of the known pools are on-stream and the Russians seem to believe that they can hold output at Samotlor at 155 mn tons a year for some time yet.

A report by Tass (3rd October 1981) announcing the recovery of 1,000 million tons at Samotlor since 1969 claims that only 6 years will be needed for the deposit to yield its next 1,000 mn tons. This may confirm the belief that the deposit holds considerably more oil than the 2,000 mn tons initially estimated. Research suggests that its recoverable reserves may exceed 3,500 mn tons, although some of this has a high slate content. Consequently, Samotlor is now being converted to gaslight so as to overcome the problem of salt deposition on submersible pumps. The first high-pressure compressor station was commissioned in 1981, 61 mks from Nizhnevartovsk, and over the period 1981-1983, 14 compressor stations will be built, 1,000 kms of gas pipelines will be laid, 250 computer-operated distribution centers will be set up and 1,500 wells will be converted to gaslift. By 1985, gaslift wells will produce 45-50 mn tons a year, or 30% of Samotlor's rated output.

The Fedorovsk deposit was discovered in 1972, and brought rapidly on-stream in March 1973. It is part of the Surgut field, and Fedorovskneft NGDU was created in 1977 when output began to grow rapidly. In 1981 it produced over 30 mn tons and ranked second after Samotlor. It is expected to peak in 1982, and the installation of 3.5 mn tons/year treatment plants indicate that the Soviets expect peak output to be maintained for at least 10 years.

Siberia's third largest producing deposit is Mamontovo, which belongs to the Yuganskneftegaz Association. It was discovered in 1965 and began producing in 1970. It yielded more than 24 mn tons in 1981 and output should increase in the future because appraisal drilling has extended the boundaries and the number of pay-zones of the deposit.

One of the largest Siberian deposits in the future will be Kholmogorsk, situated 260 kms north-east of Surgut, and connected with the trunk pipeline system by an 820 mm diameter pipeline of 25 mn tons a year capacity. The importance of Khol-mogorsk is such that a new Association, Noyabrskneftegaz, has been set up to work it. The deposit was discovered in 1973 and began producing in 1976. It lies astride the Surgut-Urengoi railway, and is served by the oil workers settlement of Noyabrskaya with a current population of 20,000. A new town, Khanto, with an even-tual population of 80,000 is to be built 11 kms north of Noyabrskaya to develop the deposit.

Drilling crews from Belorussia, Saratov and Bashkiria have been helping with the development of Kholmogorsk, and in 1981 it produced 4.2 mn tons. It is slated for rapid growth in the 1980s and, together with the nearby deposits of Karamovsk, Muravlenkovsk and Sutorminsk, it is planned to produce up to 40 mn tons in 1985. A new pipeline, running for 1,530 kms from Kholmogorsk through Surgut and Perm to Kuibyshev will be built over 1982-1984.

By the beginning of 1976, 18 deposits were being worked in Tyumen oblast, although nine had peaked and eight were approaching maximal production with drilling taking place in the peripheral and less productive parts. While Samotlor had accounted for 73 per cent of the annual increase in production over 1971-1975, its contribution was expected to fall to 33 per cent over 1976-1980. Over this latter period, a substantial share of new oil came from 22 new deposits, most of which were much smaller and at a greater distance from established population centers and transport routes than the existing deposits. They included four in 1980 - Nonyegansk, Sredne Bal-yk, Karamovsk and Yaunlorsk. The last named deposit has been developed at a rate unprecedented in the Soviet Union. It came on-stream in Ocotber 1980, and over 220,000 metres were drilled in 1981. Some of Siberia's most highly-skilled drilling crews, such as those lead by Volovodov, Manakov and Shakyurov were directed there.

Karamovsk is the most northerly oil deposit being worked, and is located 300 kms north of Surgut. It gave 0.2 mn tons in 1981, but output should rise to 8 mn tons in 1985. Over 1981-1985, 35 new deposits will be exploited, including 13 in the Nizhnevartovsk field. In 1981, five deposits were commissioned including Teplovsk (for which a special pipeline was built to the Ust Balyk gas refinery to carry the deposit's highly gaseous oil), Sutorminsk and three deposits in the new Krasnoleninsk field 250 kms west of Surgut. Muravlenkovsk should have started up in 1981 but was delayed by the failure to build the pipeline connecting it with Kholmogorsk on time.

The development of many small fields with yields far lower than those at Samotlor has resulted in a large rise in drilling requirements. Development drilling plans have usually been underfulfilled, although oil production plans have been met. This is because the usual response to drilling problems has been to step up the drilling rate at Samotlor where yields are higher and conditions easier, and where some crews drill over 100,000 metres a year.

The Tenth FYP called for 30 mn metres of development drilling by Oil Ministry crews in Tyumen oblast, but only 26 mn metres were drilled. It was initially planned to drill 80 mn metres over the current FYP period, including 34 mn metres in the Nizhnevartovsk field, but this has been scaled down, since the decision to go for gas rather than oil, to 73 mn metres. The Soviets hope to raise the drilling rate to 18 mn metres a year by 1985, more than is currently achieved throughout the country as a whole, and this should prove feasible if the 1982 plan of 13.6 mn metres is met.

Table 5

Development Drilling in Tyumen Oblast, mn metres

Field	1978	1979	1980	1981
Nizhnevartovsk	2.2	2.9	4.3	5.5
Surgut	1.3	1.7	1.9	2.8
Yugansk	1.1	1.3	1.4	1.8
Shaim	0.2	0.2	0.3	0.3
Total	4.8	6.1	7.9	10.4

Source: Authors estimates based on a wide variety of sources, and confirmed by a private communication.

Drilling speeds in Tyumen oblast are high by world standards, and new records for the Soviet Union are being continuously created. No sooner had Agafonov's crew drilled a 2,385 metre well at Fedorovsk at a commercial speed of 10,439 metres a month than Shukyurov's crew drilled a well in 5 days at Yaunlorsk at a spped of 12,580 metres a month. Speeds of over 10,600 metres a month have been regularly achieved by Boiko's crew of Yuganskneftegaz.

The total number of new development wells introduced each year has grown from 355 in 1970 to 1,016 in 1975 and 2,500 in 1980. By the end of 1981, Tyumen oblast was said to have over 14,000 oil wells, although this figure probably includes water injection wells, and the 11th FYP expects 25,000 oil, gas and water injection wells to be drilled over 1981-1985.

During the Ninth FYP, the large number of new wells being drilled at Samotlor resulted in the average yield of new wells remaining at the high level of over 60,000 tons a year, while the average yield of old wells grew to over 45,000 tons a year.

Table 6

Well Yields in Tyumen Oblast

	1971	1975	1979[a]
Well yields (tons/year):			
total	40,900	47,400	35,900
new	56,200	61,300	50,000
old	36,700	45,500	33,300
Number of wells:			
total	1,004	3,016	7,650
new (mid year average)	377	926	1,958
old	542	1,895	5,293
not working	85	195	400
Volume of oil (mn tons):			
total	41.1	143.0	274.2
new	21.2	56.8	97.9
old	19.9	86.2	176.3

a Estimated

Source: Calculated from Ekonomika Neftyanoi Promyshlennosti 1976 No. 8.

With the smaller deposits accounting for a larger share of total output in the 1980s the yields of new wells should fall dramatically, and this will tend to drag down the average yield of old wells as time goes by. Chapter 10 gives an indication of how well-yields are expected to fall during the period to 1990, and shows how the increase in drilling will more than compensate for declining well-yields.

The comments of V Filanovskii[1] regarding well-yields have attracted attention in the West. He claims that the daily yields of new wells in Western Siberia have fallen from 162 tons/day in 1975 to only 90.6 tons/day in 1978. This means that the 1,665 new wells in 1978 gave only 55.1 mn tons of oil, leaving 190.6 to be produced by the 4,504 old wells. This would give a daily yield of 116 tons per old well compared with 125 in 1975.

Assuming that new well-yields fell regularly between 1975 and 1978, then it can be shown (on the basis of Filanovskii's data) that the yields of wells which were already part of the "old" stock in 1975 actually increased over the three years 1975-1978. This would be most unusual: in the USSR yields usually fall rapidly during the first five years of exploitation. It must be assumed that Filanovskii's figure of 90.6 tons/day for new wells in 1978 refers to new wells being drilled in new deposits, and does not include the large number of high-yield wells still being drilled at Samotlor, Fedorovsk etc.

1 Planovoe Khozyaistvo 1980 No.2

Tomsk oblast

Tomsk oblast has 54 oil deposits, although most of them are comparatively small, and only 8 are currently producing. Oil production has more than doubled during the Tenth FYP from 5 mn tons in 1975 to 10.5 mn in 1980. It has so far come almost entirely from the Sovetskoe deposit belonging to Strezhevoineft directorate, and it is believed that Sovetskoe can sustain a level of output of 10 mn tons a year for many years. The installation of oilfield equipment began at the new Vasyugan field in 1977 and output is likely to grow rapidly in the future, although all the fields deposits are more than 3,000 metres deep. A pipeline has been built over the 180 kms from Vasyugan to Rabkino, where it joins the trunk line from Aleksandrovskoe to the Trans-Siberian oil pipeline.

The major deposit in the Vasyugan field is Pudino, and this should be producing 3 mn tons a year by 1985 out of a planned total output from Tomsk oblast of 15 mn tons. The deposit, which lies at a depth of between 4,000 and 4,500 metres, will come on-stream in 1982, and was originally expected to yield up to 10 mn tons a year by 1985. However, drilling speeds are extremely slow in spite of the use of bits specially designed by the VNIIBT institute, and Pudino's development has been scaled down. It will be one of a total of 7 new deposits to be developed in the oblast during the Eleventh FYP. These include Urmanskii, where oil will be obtained from Paleozoic formations at depths of 4,550 metres, a record for Tomsk oblast, and Pervomaisk, which came on-stream in 1981.

GAS

Gas extraction in Western Siberia began in 1963 when the Berezovo group of very small deposits located on the left bank of the lower reaches of the River Ob was opened up. The first customer was a local fish canner, but by 1966 a pipeline to carry the gas from the Punga deposit to Serov in the Urals was built, and the Berezovo group rapidly reached its projected peak of 9 to 10 bn cu m/year.

At the same time that the Berezovo group began producing, several supergiant gas-fields were discovered to the east, including the world's largest deposit of Urengoi in 1966. By 1972, the Siyanie Severa pipeline from Vuktyl (Komi republic) to the Moscow and Leningrad areas had been extended eastwards to Medvezhe, and the installation of the first gas preparation plant at Medvezhe permitted the deposit to begin producing.

Output grew very rapidly, and in 1975 Medvezhe produced 30 bn cu m from a Tyumen oblast total of 35.7 bn cu m.

By 1976, when Medvezhe produced nearly 41 bn cu m from a Tyumen total of 47.7 bn cu m the Siyanie Severa pipeline had been doubled up, and a third string designed to branch off at Peregrebnoe and take gas southwards to Nizhnyaya Tura and the Urals was under construction. The drilling of wells at Urengoi and Vyngapur began in 1976 in preparation for the eastwards extension of Siyanie Severa and a new pipeline running from Urengoi in a south-westerly direction to Chelyabinsk, although it was not expected that gas extraction would begin until 1978, when 15 bn cu m from Urengoi and 4 bn cu m from Vyngapur would be produced.

In 1977, the seventh gas preparation unit, of 10 bn cu m annual capacity, was installed at Medvezhe, and the completion of the eighth enabled the deposit to reach its projected peak of 65 bn cu m. The preparation of Urengoi proceeded with the creation of a settlement to work the deposit together with a mobile power

station. The settlement was initially called Yageinyi, but subsequently changed
its name to Novyi Urengoi. Appraisal drilling at Urengoi continued to reveal
more payzones, some comparatively shallow at 1,000 metres, and as many as
twelve containing condensate at depths lower than 3,000 metres. The deposit was
also shown to cover a much larger area than previously thought, with its length
put at 167 km and its breadth at 25 to 35 km. Its proved plus probable reserves
were put at 10,000 bn cu m at the end of 1980 according to the head of the
Urengoigazodobycha Association which is responsible for working it.

During the winter of 1977/1978 the Siyanie Severa pipeline was extended eastwards
yet again, this time to Urengoi. By April 1978 the first gas preparation unit
had been installed, enabling Urengoi to begin producing on April 22, 1978, two
months ahead of schedule. However, the second unit was installed two months behind
schedule in November and this meant that the plan target of 15 bn cu m could
not be reached, with only 10.8 bn cu m being produced. Moreover, the Vyngapur
deposit, which was supposed to start up in mid 1978 and send 4 bn cu m through
the Urengoi-Chelyabinsk pipeline by the end of the year, did not come on-stream
until the end of 1978, and only produced 1 bn cu m during that year. Consequently,
the plan for Tyumen oblast of 88.5 bn cu m was underfulfilled by 3.5 bn cu m.
Thanks to the installation of the ninth preparation unit, Medvezhe managed to
overfulfill its annual plan of 65 bn cu m and account for half of the shortfall by
the other two deposits.

The plan for 1979 was set at 115 bn cu m with Urengoi expected to produce 30 bn
cu m, Vyngapur 14 bn cu m and the Nadym fields (i.e. Medvezhe and the others)
71 bn cu m, and 10.3 bn cu m of casinghead gas were to be produced. The third
preparation unit was installed at Urengoi in November 1979 and the construction
of the fourth unit began. Once again, plan targets were slightly underfulfilled;
Tyumen oblast produced 111 bn cu m with Urengoi falling short by 6 bn cu m
and Medvezhe managing to make up only 2 bn cu m of the deficit.

During 1980, the fourth preparation unit at Urengoi started up, and the fifth,
which was the first with a capacity of 15 bn cu m a year, was introduced before
the end of the year, bringing Urengoi's capacity to 73 bn cu m a year. In
October 1981, the 6th unit was installed, 9 months behind schedule, and the number
of operating wells at Urengoi passed 170. However, the field again failed to
meet its plan target of 88.6 bn cu m in 1981, giving only 82 bn, and it was left
to the Nadym deposits to make up the difference.

The Eleventh FYP has set a target of 357 bn cu m for 1985, including 250-270
bn at Urengoi, and another 15 preparation plants will be installed at the deposit.
This will bring the total to 21 with an aggregate capacity of nearly 300 bn cu
m a year, suggesting that production at Urengoi will continue growing beyond
1985. The Eleventh plan contains little mention of the Yamburg deposit, which
was originally intended to be the source of the Yamal gas pipeline to Western
Europe. In fact its development has been delayed in favour of the faster expansion of Urengoi, and construction work is due to begin in mid-1982, with drilling
beginning in 1984. The first tractor convoy arrived at Yamburg from Nadym in
January 1982.

Table 7

Production of gas in Tyumen oblast, bn cu m

	1975	1979	1980	1981	1985 Plan
Medvezhe	30.0	69.5	71	72	62
Vyngapur	-	14[a]	19	20	20
Urengoi	-	24[b]	50[e]	82	250
Others	3.5	3.5	2	2	-
Total natural gas	33.5	111.0[c]	144[f]	176[g]	332
Casinghead gas	2.2	10.5[d]	12	14.3[h]	25
Total gas	35.7	122.4	156	190	357[i]

Sources: a Tyumen radio 29/12/79. b Tyumen radio 16/1/80. c Tyumen radio 3/1/80. d Izvestia 12/3/80. e Moscow radio 8/1/81. f Moscow radio 4/1/81. g Tass 25/1/82. h estimated from Tyumen radio 4/11/81. i Moscow radio 28/1/82.

The biggest single problem facing the Western Siberian gas workers at the moment is the lack of an infrastructure at Urengoi. There is no shortage of gas in Western Siberia, and the 10th five year plan's exploration target of 7,000 bn cu m of new reserves was met in 3½ years. A total of 10,500 bn cu m of new gas was found over 1976-1980. The drilling conditions are not difficult, and the plans are not very demanding compared with those set for the oil workers. The drilling directorates of Nadymgazprom, Urengoigazprom and Vyngapurgaz were expected to drill only 530 wells of 620,000 metres during the four years 1977-1980, while the Oil Ministry was asked to drill 9 mn metres during 1980 alone. The gas wells have an average depth of only 1,223 metres, and the massive well-yields give huge returns per rouble of capital employed. By 1985, it is planned to drill one million metres a year in the Tyumen gas fields.

The problem is entirely one of a poor infrastructure leading to labour shortages and a slow rate of gasfield construction. At first, Novyi Urengoi grew rapidly after work started in the winter of 1973, with a helicopter pad, a small power station and a sports centre being built, and the first street (Ulitsa Optimistov-Street of Optimists) being laid with five-storey buildings erected from panels made in Nadym. But then a lengthy argument arose over whether to make Novyi Urengoi or Tikhii the centre for the gasfield, and the pace of construction slackened, to the extent that by 1979 the town was 100,000 sq m short of housing, and 8,000 of its 20,000 inhabitants were living in temporary accommodations. There were no polyclinics, cinemas or a Palace of Culture, and during the four years 1977-1980, less than half the planned number of shops and only 64 per cent of the planned canteen facilities were built.

Planning failures must carry much of the blame for the gas preparation plants, without which the gas cannot be transported, being built behind schedule. There is no logical plan for the sequence of construction of individual objects like preparation plants, compressor stations, pipelines etc., and Minneftegazstroi has likened itself to a fire brigade, "going where it burns". Consequently, there is a lack of coordination. When the Yagenetskaya compressor station was completed, it remained idle because insufficient gas was being sent along the Urengoi-Chelyabinsk pipeline to justify its operation. The third preparation plant started up late, but even then only one of the eight well clusters designed to serve it, had been completed.

The failure to complete the 500 kv transmission line from Surgut power station to Urengoi until February 1982 meant that 700 workers were needed to operate 20 dwarf power stations. Only a fraction of the necessary length of roads has been laid because the Ministry of Transport Construction (Mintransstroi) has only recently begun to build bases for its workers. The Ministry of Gas tends to receive all the blame for these problems, but it must be remembered that it has enough problems of its own with its drilling targets invariably exceeding its capacity. Workshops for the repair of drilling equipment, casing pipe etc. have not been built yet, although they were planned for several years ago.

Another reason for the delays in drilling at Urengoi is said to be the failure to move rigs from Medvezhe and Vyngapur where drilling is virtually finished. This poor organisation of work also leads to accidents and blowouts at wells.

Other deposits are not without their problems. Medvezhe has been able to make up for a part of the shortfall at Urengoi, but its second stage of development, during which pumping stations are installed for when formation pressures begin to fall, is said to be proceeding too slowly. Vyngapur, on the other hand, is developing well after a delayed start up, with its 100 wells being drilled ahead of schedule.

The Messoyakh and Soleninsk deposits to the east of the Yenisei River are located in Krasnoyarsk krai, Eastern Siberia region. They deliver over 4 bn cu m a year through a 3-string pipeline to the important mining centre of Norilsk.

A special Council of Ministers committee has been set up to oversee the development of the Western Siberian oil and gas fields. Headed by Deputy Chairman of the Council of Ministers V.E. Dymshits and including Tyumen obkom chairman Bogomyakov and Gosplan deputy chairman Lalayants, it has the power to cut through red tape and direct necessary resources to Western Siberia. Its most important task is to ensure that the plans to produce 1 mn tons of oil a day by April 1984, and 1 bn cu m of gas a day by the first quarter of 1985 are fulfilled.

URALS-VOLGA OIL

The Urals-Volga was the most important oil-producing region in the USSR as recently as 1977. It reached a peak in 1975, and has since been declining according to plan. It is still sufficiently important to merit a massive expenditure of money, men and materials to try and slow down its decline by opening smaller and deeper deposits, and prolonging the life of existing deposits by the use of tertiary recovery methods.

The Eleventh FYP expects oil production by the RSFSR (Russian Republic) to reach 570 mn tons in 1985, and this suggests that output by the Urals-Volga region will fall from 193 mn tons in 1980 to 125 mn tons in 1985.

Table 8

Oil Production in the Urals-Volga region, mn tons

	1975	1979	1980	1981
Tatar republic	103.7	85.8	83	75.2
Bashkir republic	40.3	39.7	40.3	40.0
Kuibyshev	34.8	25	24	23.1
Perm	22.3	24	21	13.9
Orenburg	13.9	12	10	9.5
Udmurt republic	3.7	7.4	8.5	8.8
Volgograd & Saratov	8	7	6	6
Total	226.7	201	193	176.5

Tatar republic

Production in the Tatar republic reached its peak in 1975 at nearly 104 mn tons. It has since fallen sharply, although the Tenth FYP target of 83 mn tons in 1980 was fulfilled and the annual target for 1981 of 75 mn tons was slightly overfulfilled. Since production began in 1948, the Tatar republic has produced over 2,000 mn tons of oil, and it is expected that it can yield another 1,000 mn tons. The Eleventh FYP calls for the production of 300 mn tons over the five years 1981-1985, which implies an output of only 45 mn tons in 1985.

This production decline is taking place in spite of the drilling of a substantial number of new wells. During the period 1976-1980, 6,000 new wells were drilled and the plan to drill a total of 8.5 mn metres at a constant rate of 1.7 mn metres a year was fulfilled.

The region is still dominated by the supergiant Romashkino, discovered in 1948. But as the declining output from a constant drilling metrage suggests, a larger share of total production is coming from smaller fields with lower well-yields, mainly in the Almetyevsk, Leninogorsk and Kama areas. A host of tiny deposits are being discovered with the volume of exploratory drilling increasing from year to year; in 1981 a record of over 1 mn metres was achieved. The newest of the 12 oil production directorates, Murlatneft, has been set the task of working the numerous small deposits in the south of the republic. It pays to work these deposits, in spite of the high production costs, because the proximity of the Ufa and Kuibyshev refinery complexes leads to fairly low delivered costs of crude.

Bashkir republic

Having reached a peak of over 45 mn tons in 1967, production has since stabilised at 39 to 40 mn tons a year. This has come as a pleasant surprise to the planners as production is claimed to have been 9.7 mn tons more than planned during the Tenth FYP.

Unlike the Tatar republic, Bashkiria has not been dominated by a single field since the decline of Tuimazy to 4 mn tons a year, at which level the flow has been stabilised. Bolshoi Arlan has been producing around 8 mn tons a year since 1976 but at this moment most of the republic's oil is coming from 40 small deposits. More of these are still being discovered at a healthy rate - 52 have been found during the last 5 years - in spite of the claim that 95 per cent of the republic has been explored. Most of the new finds are at great depths, and wells of 5,500 metres at the Kabakovo deposit and 7,000 metres at Kulgunino have been drilled.

The most promising recent discovery is Naratovsk, south of Ufa. Discovered in 1978, it was brought on-stream in 1979, and eventually over 200 wells will be drilled. The influence of these many small deposits can be seen on the drilling plans, which called for the drilling of 8.5 mn metres over 1976-1980. This was the same as for the Tatar republic, while total oil production over the period was less than half that of the Tatar republic.

Kuibyshev oblast

Since oil was discovered in Kuibyshev oblast in the late 1930's, over 700 mn tons have been produced. Output peaked at 35.6 mn tons in 1971, and has been falling steadily ever since, as foreseen by the Ministry of Oil, to little over 23 mn tons in 1981. Exploration is continuing, with 37 new deposits discovered during 1976-1980; six of these have begun producing oil, including the large Nikonovo deposit. The use of modern equipment and tertiary recovery is slowing the decline somewhat, but Kuibyshev cannot be regarded as a major producer in the future.

Perm oblast

Although Perm oblast was the scene of the first oil strike in the Urals-Volga region in 1929, it is only since the mid-1960's that production has reached appreciable levels. Targets have been over-ambitious in the past, and the Tenth FYP target of up to 30 mn tons in 1980 was badly underfulfilled. This is in spite of the discovery and rapid exploitation of significant finds during the last five years at Blagoveshkoe near Chaikovskii and two deposits near Berezniki. The latter are of sufficient importance to justify the construction of a 160-kilometre pipeline connecting them with the Perm refinery. However, problems with heavy oil and inefficient waterflooding in the rest of the oblast caused a sharp decline in 1981 oil production to less than 14 mn tons.

Other areas of the Ural-Volga region

The other areas produce comparatively small amounts of oil. Production has been declining steadily everywhere except the Udmurt republic, where it has been growing by 1 mn tons a year throughout the 1976-1980 period and plan targets have been consistently overfulfilled. The proximity of the Druzhba pipeline, into which all Udmurt oil flows, has rendered the republic's nine small deposits economical to work.

During 1976-1980, 13 oil deposits were found, and nine of these are slated to begin producing over 1981-1985, enabling the republic to produce 45 mn tons at a constant rate of 9 mn tons a year. In 1981, three new deposits were commissioned, including Gremikhinsk, 40 kms north of Izhevsk, where steaming is being employed to recover heavy oil from depths of 1,190 metres.

CASPIAN OIL

This section considers oil production by the Azerbaidzhan, Kazakh and Turkmen republics, and from the Caspian Sea, as all these producing regions face similar problems. The oilfields of the Caspian Sea and its shores have so far proved a major disappointment. This can be judged from the fact that the plan target of 65 mn tons for 1980 compared with an actual output of about 44 mn tons, and this shortfall of 21 mn tons compares with a shortfall for the country as a whole of 37 mn tons in 1980.

Table 9

Oil production in the Caspian region, mn tons

	1975	1979	1980	1981
Azerbaidzhan	17.2	14	15[b]	15
Kazakh republic	23.9	18	18.7[a]	19.1[c]
Turkmen republic	15.6	11	7.5	7
Total	56.7	43	41	41

a Ekonomicheskaya Gazeta No. 14, 1981 b estimated from Moscow radio 15/9/80
c Moscow radio 28/12/81.

Sources: statistical yearbooks of the respective republics

Kazakhstan

Kazakhstan has three oilfields, of which the largest is Mangyshlak, consisting
of two major deposits, Uzen and Zhetybai. Production has declined from 20 mn tons
in 1975 to only 14 mn tons in 1979, when the annual plan target was underfulfilled.
The decline is due to difficulties in tapping the field's heavy oil. Extensive
waterflooding has been employed, and 20 water-pumping stations are now operating,
although some Western analysts believe that waterflooding has been too low with
well-yield below expectations. Gaslift is now being used as well as other
enhanced recovery methods (see Chapter 9).

Although Mangyshlak has given more than 200 mn tons so far, output is likely to
decline in the future. The Soviets are leaving the heavy oil problem for the time
being, preferring to develop the easier deposits like Zhilanda, which is Mangyshlak's
deepest operating deposit with its first well giving gas from 3,412 metres.

The second largest Kazakh field is the Emba, situated on the northern shore of
the Caspian. After being worked for many years, it gave only 3.8 mn tons in 1978,
and the Tenth FYP target of 5 mn tons in 1980 was not reached. However, output has
been rising slightly, mainly from the new Tortai deposit, where wells drilled to
3,000 metres are giving very high yields of 450 tons a day.

The third field is situated on the Buzachi Peninsula, north of Mangyshlak,
and was established in 1977 with reserves of more than 150 mn tons. It is being
rapidly developed so as to hold up Kazakh production until the heavy oil problems
at Mangyshlak are solved. While the oil is mostly heavy, it lies at depths of
only 250 to 800 metres, and can be extracted comparatively cheaply by the drilling
of at least 14,000 wells. There are three major deposits; Karazhanbas and
Severobuzachinsk were to have come on-stream in 1978, and Kalamkas began producing
in 1979. Karazhanbas was developed two years late in 1980, and Severobuzachinsk
was commissioned in April 1981. The oil is carried by a 267 km pipeline running
from Kalamkas through Karazhanbas to the port of Shevchenko and from here it is
shipped to the Baku refineries. As output grows, some oil will be despatched
through the Uzen-Gurex-Kuibyshev pipeline which is specially heated to keep
the temperature of the heavy oil above its congealing point.

Kalamkas deposit is liable to flooding by the Caspian Sea, and so a 32 km dyke
has been built to protect it. Sand invasion and paraffin deposition are also
hampering development, and it is not yet certain that progress will be any easier
at Buzachi than at Mangyshlak.

Output has grown to 1.5 mn tons in 1981, and 2 mn tons is planned in 1982. The field, which is being administered by the new Komsomolskneft Association, has been set a target of 5 mn tons in 1985. Kazakhstan as a whole is planned to produce 25 mn tons, and some of this will come from a new oilfield which will be established in the Caspian Depression in the southern part of Aktyubinsk oblast. It will centre on the Zhanazhol deposit which the Oktyabrskneft Association will bring on-stream in 1982. By the end of 1981, 30 exploratory wells had yielded large quantities of light oil from two pay-zones lying between 3,000 and 4,500 metres deep, and four other sub-salt deposits had been found in the Caspian Depression.

Azerbaidzhan

Azerbaidzhan was once the most important oil-producing region in the USSR, and indeed the world, but oil production has been declining since it peaked in 1966 at 21.7 mn tons. Production has dropped as the very old on-shore deposits of the Apsheron Peninsula have become exhausted. Nowadays 60 per cent of oil and 90 per cent of gas comes from off-shore deposits.

It is hoped that oil production will begin to rise again during the 1980s as newly discovered deposits in the Caspian Sea are developed. This depends on the speed with which the Russians can augment their fleet of modern jack-up and semi-submersible rigs, because the most promising deposits lie in up to 200 metres of water.

Caspian oil extraction is organised from the city of Neftyanye Kamny which has been built on fixed platforms 40 kms from the shore on the oil deposit of the same name. At its peak in 1965, Neftyanye Kamny produced nearly 9 mn tons a year, but output fell to 3.5 mn tons in 1979 before rising to 4.1 mn in 1980. It is believed that the development of a new north-east section and an increased gas-lift capacity will enable output to rise further in the future. The use of gaslift is enabling abandoned wells to be reactivated. To the south-east, the Neftyanye Kamny-2 extension (4,700 metres deep) is being opened up.

The anticipated rise in Caspian output has prompted the extension of the facilities at Neftyanye Kamny, including the construction of a large new oil collection and treatment centre. Much optimism stems from the discovery of the "28th April" deposit in August 1979. It lies 15 kms east of Neftyanye Kamny in 84-140 metres of water, and the first well gave a flow of 250 tons a day from 3,500 metres. Within six months, an 11 kms pipeline had been laid connecting the deposit with the collection centre, from which the oil was carried to Baku by tanker. The pipeline was laid by the special pipelaying ship, the Suleiman Vezirov, at a rate of 800 metres a day, and was extended further to Zhiloi Island by September 1980 and Baku by July 1981. A second well was completed by Ismailov's crew in December 1980, and "28th April" has been confirmed as being the most promising deposit in the Caspian. Another eight inclined wells will be drilled from the same platform, and the semi-submersible "Kaspmorneft" has begun drilling the deposits third well with a rated depth of 4,500 metres in 140 metres of water.

A fixed platform is being erected in two sections at "28th April" in 110 metres of water. The sections, each weighing 2,000 tons, stand 116 metres high and the platform will carry two rigs so that two wells can be drilled simultaneously. A total of 12 wells will be drilled from the platform, and during 1981-1985, nine such platforms will be erected at the deposit.

Other Caspian deposits expected to contribute to the production increase of the 1980s are Yuzhnaya-2, Bakhar, Bulla-More and Banka Andreyeva. At Yuzhnaya-2, large pools of oil and gas are expected at depths of up to 5,300 metres and the jack-up "Baku" has drilled to this depth, but no major finds have been reported. At Bakhar, wells of up to 5,479 metres are being drilled, and oil has flowed from

5,200 metres. New wells at Bulla More are yielding 0.7 mn cu m of gas and 200 tons of condensate a day from its 7th horizon at a depth of 6,055 metres, and the number of wells is expected to rise from 15 to 25 over the next few years, all 6,000 metres deep. At the moment, these deposits are more important for their gas, with Bakhar, Duvannyi and Bulla More giving most of Azerbaidzhans 13.8 bn cu m in 1981.

On-shore drilling at Azerbaidzhan will be concentrated on the new deposit of Muradkhanli. Situated in the Kura depression, it was discovered in 1971 and began producing from wells of 4,000 metres in 1977. Some ten rigs are currently operating, but new wells must be drilled to 5,600 metres, and development is slow due to delays in building roads through the swamps.

During the Tenth FYP period, more than 2,000 mn roubles were invested in the oil and gas industries of Azerbaidzhan, compared with 1,400 mn during 1971-1975, although some important construction targets have been badly underfulfilled. Poor labour organisation is said to be the principal problem.

Turkmen republic

Most Turkmen oil comes from the on-shore deposits of Kotur Tepe, Barsa Gelmes and Nebit Dag, with 0.4 mn tons a year from the Cheleken Peninsula and a small amount from the off-shore deposits of the Caspian Sea.

On-shore production has been falling, and the republic produced only 55 mn tons over 1976-1980, or 14 mn tons less than planned. This is because of problems with heavy oil, paraffin deposition, sand invasion and anomalously high formation pressures which cause very high initial yields (such as 800 tons a day at a new Kotur Tepe well) but lead to blowouts and breakdowns. Most of the oil output during the Tenth FYP period was to have come from the south-east extension of the Kotur Tepe deposit, but in addition to the problems mentioned above, the plan for water-flooding was not fulfilled, and the Soviets have decided to install gaslift. At other deposits, waterflooding was carried out inefficiently and some wells were ruined. However the Turkmenneft Association plans to drill 50% more wells over 1981-1985 than during 1976-1980, and during 1981 the oilflow at Barsa Gelmes is said to have doubled to 2.2 mn tons thanks to the installation of new waterflood pumping units and gaslift compressors using gas from Burun delivered through a 20 kms pipeline. In August 1981, the first compressor station using casinghead gas from Barsa Gelmes came into service, and Turkmenneft plans to install three more compressor stations, enabling all the casinghead gas to be used.

The search for new oil is being concentrated on depths of more than 5,000 metres. By April 1980, 52 wells had been drilled to this depth, with more than half giving commercial flows of oil. In recent years, a number of promising new deposits such as Gograndag, Kuydzhik and Burun have been discovered in Western Turkmenia. Drilling work is being speeded up by the use of new modular rigs, but at places like Gograndag the lack of roads is delaying their delivery.

It is likely to be several years before the decline in on-shore production can be halted, but things look brighter in the Caspian Sea, where seven deposits have been discovered of which two are producing. They are administered by the Kaspmorneftegazprom Association, which is responsible for all Caspian Sea oil production.

Exploration work began at Banka Zhdanova in 1968 and this deposit has since become the main base for further development in the eastern Caspian, with a forest of platforms stretching for 10 kms. The first wells yielded 400 tons of condensate a day, and over the period 1971 to 1975 some 3 mn tons were obtained. New wells are being drilled to depths of up to 5,000 metres but pressures are very high, and for a lengthy period the wells could not be operated because of a shortage of high strength well-head equipment. In June 1981, another platform (the 60th on the eastern side of the Caspian) was completed on Banka Zhdanova and 3 wells will be drilled from it.

Banka Zhdanova is connected to the mainland by a pipeline, which has now been extended by 17 km to another important deposit, Banka Lama. Here, four rigs are operating, and they have drilled 15 wells which yield up to 300 tons of condensate a day. Another pipeline is being built over the 40 km from Lam to the shore; from here the oil is piped to the Kotur Tepe deposit where the gas is separated and despatched along the Central Asia-Moscow pipeline, and the oil is piped to the refinery at Krasnovodsk.

Eventually the Caspian pipeline system will be extended westwards to the Livanov group of deposits where the Khazar jack-up has drilled eight oil-producing wells. The deposits are said to be very promising. High hopes are also entertained for the Zapadnoe Erdekli deposit where drilling began in 1978. Geologists believe that deposits on the Turkmen side of the Caspian can prove as productive as those on the Azerbaidzhan side and the arrival of new rigs able to drill to 6,000 metres of water should enable a sizeable increase in output to be obtained from this region during the 1980s.

CENTRAL ASIAN GAS

Gas has been produced in Central Asia since before the last war, but output only became significant during the mid 1960s when the giant Gazil deposit in the Uzbek republic was developed. It grew from 0.7 bn cu m in 1960 to 17.9 in 1965, 46 in 1970 and 89.7 bn cu m in 1975. With Gazli and the other Uzbek deposits stabilising at 32 to 37 bn cu m during 1970-1975, the new Turkmen fields were responsible for almost all the impressive increase in gas production during the early 1970s.

Table 10

Gas production in Central Asia, bn cu m

	1975	1979	1980
Uzbek republic	37	36	35
Turkmen republic	52	68	71
Total Central Asia	90	104	106

Sources: estimated from Gazovaya Promyshlennost and Neftyanoe Khozyaistvo

The Turkmengazprom Association accounts for most of the output of the Turkmen republic, with Kaspmorneftegazprom producing the rest from the Caspian Sea and Western Turkmenia. Turkmengazprom consists of two associations, Shatlykgazodobycha and Achakgazodobycha; in 1980 it produced 63.2 bn cu m of gas, with Shatlykgaz-odobycha accounting for over 40 bn cu m made up of 36 bn from the supergiant Shatlyk deposit and 4.6 bn from smaller deposits. Shatlyk is one of the ten largest gas deposits in the world, and its initial reserves of 1,000 bn cu m could prove to be very much larger with the recent discovery of new payzones. Achakgazodobycha produced 22.6 bn cu m in 1980, mostly from the Achak, Kirpichli, Naip and Gugurtli deposits.

Table 11

Gas production in the Turkmen republic, bn cu m

Turkmengazprom	54.5	60.2	61.4	63.2[b]
Shatlykgazodobycha	30.7	36.5[c]	38.4[d]	40.6
Achakgazodobycha	23.8	23.7[e]	23.0	22.6
Western Turkmen & Caspian	8.1	7.0	6.6	7.3
Total	62.6	67.2	68.0	70.5[f]

a Moscow radio 10/1/80 b forecast by Turkmenskaya Iskra 19/11/80
c Ashkabad radio 18/1/78 d Turkmenskaya Iskra 25/2/79 e estimate from
Ashkhabad radio 21/9/77 f Turkmenskaya Iskra 20/1/81.

Output by the Achak association fell throughout the Tenth FYP period, primarily
because of the decline of the Achak deposit, although this situation was reversed
in 1980 when the new Kirpichli deposit with a rated output of 8 bn cu m a year
came on-stream. In addition, several other small deposits like Beurdeshik and
Severo-Balkui have been developed and it is likely that the association will be
producing at about 25 bn cu m a year for the rest of the decade.

Production from Shatlykgazodobycha has been growing steadily as the Shatlyk
deposit has been brought up to its projected capacity of 36.5 bn cu m a year,
Bayram Ali has reached its design capacity of 4 bn cu m a year and a number of
other small deposits have begun producing.

The Eleventh FYP has set a target for the Turkmen republic of 82 bn cu m a year in
1985. The increase in output will come mainly from the recently discovered
Dauletabad-Donmez group of deposits near Serakhs on the border with Iran, with
initial reserves officially placed at more than 1,000 bn cu m. The first development wells are now being drilled, treatment plants are being built and the deposit
will come on-stream in 1983 when it is planned to produce 8 bn cu m a year.
During the Twelfth FYP period 1986-1990, the Sovetabad deposit, with reserves
currently put at 1,380 bn cu m will start producing.

The Western Turkmen gasfields do not come under the jurisdiction of the
Turkmengazprom Association. In 1980, 6 bn cu m was produced from the Okarem,
Kuydzhik and Barsa Gelmes deposits, as well as the Caspian Sea, and output is
likely to rise in the future with the development of the Kamyshldzha and Gograndag
deposits.

Production from Uzbekistan is likely to remain at the 35 bn cu m a year mark for
the next five years. The decline of Gazli and some other aging fields will be
covered by greater output of highly sulphurous gas from the Mubarek field, principally the Urtabulak, Uchkir, Yuzhno-Mubarek and Severo-Mubarek deposits. The
Mubarek treatment plant, 15 kms east of Karshi, now has a capacity of 15 bn cu
m a year and it will be extended to 25 bn cu m a year by 1985. Built jointly by
Soviet and Bulgarian workers, Mubarek also produces 460,000 tons a year of liquid
sulphur for local fertiliser production.

The first stage of the new Shurtan deposit in the Karshi Steppe with 500 bn cu m
of reserves came on-stream in 1980. The second stage started up at the end of
1981, bringing its population rate to 8 bn cu m a year. The gas is despatched
through a 400 km pipeline to the Syrdarinsk power station (3,200 MW) which was
converted from fuel oil to natural gas in 1980 and uses half the Shurtan gas
while the rest is sent on to Tashkent. By 1985, Shurtan should be producing 16
to 20 bn cu m a year. Other important deposits are Zevardy and Dengizkul-
Khauzak, which are each producing 2 bn cu m a year.

ORENBURG GAS

The Orenburg gasfield was discovered in 1966, and was initially believed to hold 1,800 bn cu m of gas. It was developed particularly rapidly because of its proximity to the important industrial regions of the Urals and the Volga and also because it was intended to become an important source of gas for export.

The presence of sulphur in the gas meant that it had to be separated out before the gas could be despatched through pipelines. Consequently, it was decided to develop the field in stages with three gas-cleaning plants, each of capacity 15 bn cu m a year, being built. The first two plants would produce clean gas for domestic use while the third would provide gas for export to Eastern Europe through the 2,750 km Soyuz pipeline.

By 1976, the first two plants were operating, and 31.8 bn cu m of gas were produced (gas also comes from other small deposits and the Orenburg oilfield). In 1977, the production plan of 33 bn cu m was overfulfilled, with 34.6 bn cu m being produced, and the construction of the third plant was under way. In 1978 it was completed, and the field's total capacity rose to 45 bn cu m a year.

In 1981, 49 bn cu m were produced, and this production rate can be maintained up to the year 2000. Although appraisal drilling is continuing, and the field is proving bigger than originally thought, there are no plans at present to increase the capacity of the processing complex.

This may be because disproportions have crept into the planning process. Housing construction plans have not been fulfilled, and the shortage of housing for geologists has affected the search for more gas. Other reasons for the failure to meet the plan for new gas reserves are insufficient turbo-drills, transport facilities and chemical reagents.

A helium plant, nine times larger than any previous Soviet helium plant, and designed to process 3 bn cu m of gas a year, came on-stream in May 1980. But Minneftegazstroi is lagging in the construction of the Orenburg-Kazan pipeline which will not start pumping the bi-product of ethane until the end of 1982, and it has been suggested that Gosplan and the Ministry of the Chemical Industry (Minchemprom) have no clear idea on what to do with the ethane. The second stage of the helium plant is consequently being delayed, and in 1980 only 29% of the planned construction work was carried out.

The 3 mn tons a year of condensate produced by the deposit is sent by pipeline to the Salavat petrochemical plant, but in 1979 the plant was unable to receive 100,000 tons which had to be pumped back into the ground. It has therefore been decided to build a condensate refinery in Orenburg for the production of auto petrol, and deliveries of condensate to Salavat and other Bashkir customers will end.

Orenburg also produces one million tons of sulphur and 0.5 mn tons of LPG a year. The experience gained in the extraction and processing of sulphurous gas is proving vital to the planned development of similar deposits like Astrakhan, Karachaganat (Kazakhstan) where Orenburggazodobycha workers are installing equipment, and Berkutovsk in the Bashkir republic.

Although the Komi republic had been producing a small amount of gas and oil for many years, it was not until the giant Vuktyl gas and gas condensate deposit was discovered in 1965 that the region became an important producer. Vuktyl came on-stream in 1968, and now gives 15 bn cu m of gas a year plus a large volume of gas condensate. A number of other gas, condensate and oil deposits were also developed, and by 1980 the republic was producing 19.0 bn cu m of gas and 18.1 mn tons of oil and condensate a year.

The plan target for 1975 has been set at 21.7 bn cu m of gas and 23 mn tons of oil and condensate. Total output of gas from Vuktyl will decline from 93.3 bn cu m over 1976-1980 to 84.8 bn over 1981-1985, although development will continue, and 30 more wells will be drilled by 1983 bringing the total to 150. The exploitation of a new gas deposit, Zapadno-Soplesskoe 80 kms from Vuktyl, began in November 1981, and it should reach its rated capacity of 2 bn cu m a year by 1985.

Table 12

Komi Republic Oil and Gas Production

	1975	1979	1980	1981
Oil & gas condensate (mn tons)	11.1[a]	18	18.1[b]	19.5
Natural gas (bn cu m)	18.5[c]	19	19	20

a Estimate from Moscow radio 29/11/76   b Ekon Gaz No. 14, 1981   c Estimate from Moscow radio 2/12/76.

The increase in oil output since 1975 stemmed from the development of two new oil deposits, Usa and Vozei, and by 1980 Usa was accounting for 75% of the republic's oil. But while production has grown steadily, it has failed to meet plan targets because of shortfalls at Usa. The northernmost oilfield in the USSR, it was connected with the small refinery at Ukhta by a pipeline completed in 1975, and then extended to the larger refinery at Yaroslavl by 1978. Production short-falls stemmed from delays in the construction of the 100 km Synya-Usa branch line of the Kotlas-Vorkuta railway. It was to have been ready in 1978, but did not reach its destination until December 1979.

Of the 79.3 mn tons of oil and condensate produced by the Komi republic over 1976-1980 (some sources say 82 mn tons), 67.5 mn tons came from Usa and Vozei. However several other small deposits have begun to produce oil during the last few years, including the promising Severo-Sabinoborsk. Development is assisted by the use of modular rigs which enables them to be erected in half the usual time. More than one million metres of wells were drilled in 1981 by Komineft and Komigazprom, and over 1981-1985 they plan to drill a total 5.5 mn metres.

Meanwhile, exploration rigs are being taken further north, and the mouth of the Pechora river is now the scene of extensive drilling activity. Several sizeable deposits of oil and gas have been discovered there. Oil has been dis-covered in several places to the north of the Komi republic, such as Varandei and Severo-Seremboi, but it is likely to be several years before these deposits are developed.

A gasfield at Vasilkovo has recently begun producing to supply the town of Naryan Mar. Casinghead gas from Usa oilfield is now being processed by the Usa gas refinery, completed in January 1980, and the dry gas will be used as a fuel for the Pechora power station now being built.

OTHER SOURCES OF OIL AND GAS

The North Caucasus region has been producing oil since it was discovered at Groznyi in the 1890s. During the 1960s, production grew rapidly to a peak of 36.3 mn tons in 1971, since then it has fallen precipitously to 24 mn tons in 1975 and further to 21 mn tons in 1979. This is in line with the Tenth FYP which foresaw an output of only 20 mn tons by the region in 1980.

The proximity of a large refining complex at Groznyi and the need to keep it operating at close on capacity has justified efforts to find and produce more oil in the region. Accordingly, wells are being sunk to great depths. During 1978, it was planned to complete 39 wells of between 5,000 and 7,500 metres deep in the Chechen-Ingush republic alone, and by the end of 1978, 100 such wells were being drilled. Most of these wells have to pass through 800 metres of salt, and take up to seven years to complete.

It is thought that the oil-bearing strata lying at a depth of 5,800 metres, of which the large new deposit of Andreyevskoe is a part, may stretch for over 200 km and ensure production by the region for a long time yet. Increases in output from the Chechen-Ingush and Dagestan republics should more than compensate for small declines in the Stavropol and Krasnodar areas, and permit output by the region as a whole to stabilise at about 20 mn tons a year.

Oil production in three western regions of the USSR, the Ukraine, Belorussia and Kaliningrad, peaked at 21 mn tons in 1975, but has since declined sharply to 13 mn tons in 1980 with the Ukraine giving 8.3 mn tons. They were expected to produce up to 19 mn tons in 1980 by the Tenth FYP, but new oil is being found at very great depths in difficult geological conditions, and Belorussian and Ukrainian drillers are better employed in Western Siberia. The Shevchenko-1 well, in the Western Ukraine, is the deepest oil-well in the USSR with a depth of 7,520 metres.

The Kaliningrad oblast has raised its annual output from 0.3 mn tons in 1975 to 1.5 mn tons in 1981, and this helps to keep the Drogobych refinery (in the western Ukraine) supplied with oil. During the current FYP, the construction of a pipeline to the Mazheikiai refinery (Lithuania) will begin, and by 1985, Kaliningrad is planned to produce 3 mn tons a year.

The Ukraine was the USSR's main gas-producing region for many years, particularly after the giant Shebelinka gasfield came on-stream in 1956. When Shebelinka began to decline in 1972, Ukrainian output was sustained by two other fields located near Shebelinka-Yefremovka and Krestishche. However it was foreseen that production would fall during the Tenth FYP period from 69 bn cu m in 1975 to 54 bn in 1980. The plan was overfulfilled with the Ukraine giving 55 bn cu m in 1980, and this level was maintained in 1981.

Large new finds are still being made in the Ukraine. An important new pay-zone was discovered at Shebelinka in 1978, and several deposits have been found in the Black Sea to the west of the Crimea. These include the giant Golitsyn field, where several development wells were drilled in 1981, and gas will start to flow when the 83 km pipeline to Chernomorskoe on the Crimea has been commissioned in 1982. In all, 20 new oil and gas deposits were found in the Ukraine over 1976-1980.

The Far Eastern island of Sakhalin is a long-standing (since 1921) producer of oil, and its importance lies in its future potential as a source of oil to cover the estimated 15 mn tons a year requirements of the Soviet Far East and for export to Japan. Production has stabilised at 2.4 mn tons a year, but will rise slightly with the further development of the Mongi field, which was discovered in 1975, and is now yielding 0.5 mn tons a year.

Interest now centres on off-shore drilling into the continental shelf north-west of Kholmsk and south-east of Okha. In December 1976, a protocol was signed in Tokyo on joint Japanese-Soviet prospecting, with Japan promising to lease a rig and provide $152.5 mn in finance. It was believed that off-shore production may rise to 10 mn tons a year, of which Japan would receive half. In 1979, the Chaivo oilfield was discovered 12 kms off the north-east coast of Sakhalin. The pilot well produced 200 tons of oil and 1 mn cu m of gas a day. It was drilled into part of the Odoptu structure which was discovered in 1971 and is now being explored by Sodeco of Tokyo. Test drilling revealed seven strata of oil and gas at depths of between 2,000 and 2,800 metres, with some of the oil-bearing strata being 35 metres thick. Another well gave pay of 300 tons a day in August 1981, and throughout the Sakhalin coastal shelf, reserves of more than 100 mn tons have been found. In all, 15 wells have been drilled.

CHAPTER 3

Transport of Oil and Gas

THE TRANSPORT PATTERN

The transport of oil and gas during the Tenth FYP period was characterised by a growing importance of pipelines. This was inevitable, given that the entire increment in oil production now comes from Western Siberia, from whence all oil is carried by pipeline.

However the absolute volume of oil and oil products has continued to grow, albeit very slowly. This is because the pipeline construction organisations are kept at full stretch in building the crude oil and gas arteries out of Western Siberia, and product pipeline construction has lagged behind its very modest plans. Some 90% of all light products and all heavy products such as fuel oil, lubes etc. are carried by rail, and oil and its products still account for 11 per cent of all rail loads.

This trend is expected to change over 1981-1985. It is proposed to transfer some rail traffic in crude oil to pipelines bringing the volume of crude carried by pipeline in 1985 to a planned 602 to 605 mn tons. This will enable the amount of crude carried by rail to decline by 7 to 10 mn tons, but an additional 42 to 45 mn tons of products will be railed raising the total volume of crude and products to 458 mn tons in 1985 compared with 423 mn in 1980.

Table 13

Transport of oil and oil products (mn tons)

Pipeline	498.3	609.0	630.2	638[a]
Rail	389.0	407.7	422.7	421.1[b]
River	39.0	38.2	41.3	
Sea	91.4	107.9	109.9	
Total	1,017.7	1,163.8	1,204.1	1,216(est)

a Pravda 24/1/82   b Tass 4/2/82

Source: Narodnoe khozyaistvo 1980

Transport by road is confined to the distribution of products over comparatively short distances from distribution bases to consumers.

While the volume of oil loads carried by rail has risen, the turnover (measured in ton-kms) has fallen significantly from 481.4 bn ton-kms in 1975 to 460.8 bn in 1980. This is because less crude is being hauled over long distances, and more products are being hauled over short distances. The average length of haul has fallen accordingly, from 1,237 kms in 1975 to 1,090 kms in 1980.

The average length of journey by pipeline, on the other hand, is growing dramatically as more and more oil is pumped the huge distance from Western Siberia to the populated western regions of the country. By 1975, a volume of 498 mn tons and a turnover of 666 bn ton-kms gave an average journey of 1,336 kms. By 1981, the respective figures had increased to 638 mn tons, 1,263 bn ton-kms and 1,979 kms.

In spite of the growing volume of oil carried by pipeline, 35 per cent of all oil freight is carried by rail compared with 38 per cent in 1975. To send oil by rail costs three times as much as by pipeline, and the delay in pipeline construction (especially product lines) costs the USSR 700 mn roubles a year.

While the total volume of oil loads hauled by rail grew by 6 per cent over 1975-1979, the volume of oil carried over short distances grew by 20 per cent. Short distance hauls by rail are particularly uneconomical, because the cost of loading and unloading is the same as for very long hauls. Product pipelines built over these short distances would pay for themselves very quickly, especially if fuel oil pipelines connected the refineries to power stations.

The problem of filling different tank-cars with different products has not been adequately solved, and sometimes traffic jams involving thousands of tank-cars take place. Many refineries lack equipment for cleaning tank-cars when a different load (e.g. petrol instead of fuel oil) has to be pumped into them. This could be solved by a rigid specialisation of tank-cars for heavy and light products, but would result in additional runs, a lower usage rate, and new problems stemming from seasonal changes in the product mix by refineries.

The average length of journey by gas has grown considerably in recent years due to the growing share of Western Siberia and Central Asia in total output. In 1980, 401 bn cu m of gas were transported by pipeline compared with 279 bn in 1975.

PIPELINE TRANSPORT

By the end of 1975, the trunk pipeline network (i.e. excluding local gas distribution pipelines) measured 155,800 kms in length, and the Tenth FYP target called for the construction of a further 53,500 kms. This was to consist of 35,000 kms of gas pipeline, 15,000 to carry crude oil and 3,500 carrying oil products. In the event, only 45,900 kms were built (86% plan fulfillment) including 32,800 kms of gas pipeline (93.7% fulfillment) and 13,100 kms of crude and product lines (70.8%). During the 11th FYP period, 45,000 kms of gas pipeline and 10,000 kms of crude and 9,000 kms of product lines are to be built. The rate of pipeline construction is planned to average 12,800 kms a year compared with 9,200 kms a year over 1976-1980. In 1981, it was proposed to lay 14,000 kms of pipeline, including 8,800 kms of gas, 3,300 kms of crude and 1,300 kms of product lines, and a further 36 oil pumping stations and 64 gas compressor stations were to be built. However, only 12,000 kms were completed, including 6,600 kms of gas and 3,100 kms of oil and product lines. The rest consisted of ammonia and ethylene pipelines. The 1982 annual plan foresees the laying of 7,500 kms of gas pipeline with 52 compressor stations.

Table 14

Soviet Oil and Gas Pipeline Network, '000 kms

Crude & oil products	56.6	67.3	69.7	72.8[a]	89[b]
Gas	99.2	124.4	132.0	138.6[a]	177
Total	155.8	191.7	201.7	211.4	266

a Tass 23/1/82 b Moscow radio 19/1/82.

Source: Naradnoe khozyaistvo 1980

All pipelines are built by organisations belonging to Minneftegazstroi. Each organisation may typically consist of 600 workers who are permanently engaged in building pipelines passing through their territory. Some organisations, such as Omsknefteprovodstroi, are often sent to work in distant parts of Western Siberia, while others like Lengazspetsstroi, an elite organisation with 4,000 workers, specialise in difficult sections such as river crossings, trestle construction etc. The organisation can work on either oil or gas pipelines as needed.

The rate at which they are expected to work obviously depends on the terrain in which they are operating and the diameter of the pipe. In spite of the fact that the installation of large-diameter pipe has grown considerably over the current plan period (12,000 kms over 1976-1980 compared with 2,700 over 1971-1975), the construction rate has accelerated, and a pipelaying organisation is expected to average 2 kms a day of 1,420 mm pipeline.

All the major trunk lines are now built from pipe of 720 mm to 1,420 mm diameter. In 1976, such pipe accounted for 62 per cent of total crude oil pipeline with pipe of 1,020 and 1,220 mm alone accounting for 25 per cent. This share should rise considerably in the future. While the widest oil pipelines are of 1,220 mm many of the new trunk gas pipelines are being built of 1,420 mm pipe. This is not produced in sufficient quantities in the USSR and has to be imported, mostly from Japan, West Germany, Italy and France.

The principal Japanese suppliers, Nippon Steel, Nippon Kokan, Sumitomo and Kawasaki, have agreed in principal to deliver one million tons a year of pipe over the 5 years 1981-1985, with Kobe Steel also supplying plates for pipes. During July 1981-March 1982, the Russians received 750,000 tons of pipe from Japan at a reputed cost of only $400 mn. This implies a price of $553 a ton compared with a computed average price of $696 a ton for 1980's deliveries. The steel companies also offered 3.5 mn tons for the Yamal pipeline, but as a result of Japanese sanctions against the Soviet Union, it seems that at least a third of that order has gone to Mannesmann and Thyssen of West Germany. However, the Japanese contracted to supply 700,000 tons for the period April-December 1982, plus a further 95,000 tons during January-March 1983.

In 1980, West German firms (mainly Mannesmann) supplied 861,000 tons of pipe at an average cost of $570 a ton. Mannesmann has sold over 7 mn tons since 1960, at a rate of 700,000 tons a year for the last few years. The latest orders are 550,000 in April 1981, and 1.2 mn tons in February 1982. Salzgitter also supplies smaller volumes, including 100,000 tons of spirally welded pipe in July 1981 worth 150 mn DM.

Austria will become an important supplier of pipe in the future. The Voest-Alpine company has contracted to supply 800,000 tons of pipes over 1982-1985 from their Kindberg factory. The deal is worth 14 bn Sch and stems from a desire on the part of the Soviets to balance their trade with Austria. France sold 285,000 tons of pipe to the Russians in 1980, and Vallourec have been selling 150,000 tons a year for several years. While the British Steel Corporation does not make 1,420 mm pipe (its widest pipe is 1,120 mm) it has supplied large quantities of plate for pipe, and other products such as oil-well casing and tubing in significant quantities.

The USSR also buys pipe from Rumania and Csechoslovakia, which delivered 180,000 tons of small diameter pipe in 1981 and also supplied equipment to the Ilyich Sttel Plant in Shdanov to produce 2½ mn tons a year of steel plate for gas pipe-lines. The Csechs contracted to supply 392,000 tons of pipe in 1982 worth 169 mn roubles, and making Csechoslovakia the third most important supplier of pipe to the Soviet Union.

Table 15

Imports of Pipe from OECD Countries, 1977-1980

	Japan	W.Germany	Italy	France	Others	Total
Volume - '000 tons						
1977	725	618	588	324	53	2,308
1978	1,126	955	506	223	99	2,909
1979	1,242	1,032	428	274	112	3,088
1980	793	861	353	285	135	2,427
Volume - mn $						
1977	362	413	307	180	39	1,301
1978	555	472	238	135	60	1,460
1979	772	588	257	197	82	1,896
1980	552	491	203	171	144	1,561

Source: OECD Foreign Trade Statistics Series C

The USSR produced 18.3 mn tons of pipes in 1981, an increase of 14.4% over 1975 when 16.0 mn tons were produced. The principal Soviet pipe plants are:

* Almetyevsk Spiral Welded Pipe Plant, which is owned by Minneftegazstroi. A new shop producing 100,000 tons a year of long-lasting pipe with a polyethylene outer covering and a phosphate paint inner coating started up in January 1981, although Gas Minister Orudzhev complained at the 26th Party Congress that gas workers were still awaiting this type of pipe, and the first batch of pipes did not emerge until January 1982. Output by the Almetyevsk plant is rated to double by 1985.

* Severskii Pipe Plant at Polevskoi near Sverdlovsk, which increased its capacity by 150,000 tons a year in 1980 with the commissioning of a new 102-220 mill.

* Novomoskovsk Piperolling Plant near Dnepropetrovsk in the Ukraine. This is the largest pipe plant in the USSR, and it increased its output by 27% over 1976-1980 with a slightly declining workforce. It has begun producing experimental batches of spirally welded double walled pipe of large diameter, and these are currently being tested at Urengoi.

* Vyksa Pipe Plant, where the serial output of laminar pipe of 1,020, 1,220 and 1,420 mm diameter will take place for the first time in the USSR from a new shop with a capacity of 1 mn tons year. The first section, of capacity 250,000 tons a year was supposed to have started up in 1981, but its opening date has been put back to mid-1982. The pipes will have 4 to 6 layers and will be able to withstand pressures of 120 ats, enabling throughput to be raised from 33 to 50 bn cu m a year for a 1,420 mm pipeline.

* Pervouralsk Novotrubnyi Plant, located 20 kms west of Sverdlovsk, has been reconstructed recently, with the annual output of its 30-10 mill raised by 30,000 tons a year.

* Khartsyzsk. This is the Soviet Union's first pipe mill to make 1,420 mm diameter pipe. Experimental production began in May 1981, and should build up to 1 mn tons a year by 1985.

* Volzhskii Pipe Plant, near Volgograd, where an extension completed in 1981 increased its daily output by over 30% to 300,000 tons a year. The plants rated capacity is now 360,000 tons a year, and this should be reached in 1982. Volzhskii is the third plant in the USSR to begin production of 1,420 mm gas pipe, with its first experimental batch despatched to Western Siberia in January 1982.

* Chelyabinsk Pipe-Rolling Plant, which is a major supplier of 1,220 mm pipe for oil pipelines.

Several other pipe mills, the largest of which is located at Sumgait in Azerbaidzhan, specialise in the production of drill and casing pipe for oil and gas wells.

More than half of the steel plate used for Soviet-made oil and gas pipe comes from the Novolipetsk Steel Combine, where a recent reconstruction has enabled its annual level of output to be raised by 500,000 tons. During the last FYP period, plate making capacity was installed at the "Ilyich" and "Azovstal" steel plants at Shdanov, and these should become important suppliers to the piperolling sector during the next five years.

Pipeline construction problems

Pipelines are very expensive to build, especially in the difficult conditions of Western Siberia, where construction costs average 100% more than in the rest of the country. Normally, a 1,220 mm oil pipeline costs 260,800 roubles per kilometre, including 171,900 for trunk pipeline and 82,700 roubles for the pumping stations, but the 5,585 kms of trunk pipeline built in Western Siberia over 1976-1980 by the Glavsibtruboprovodstroi Association cost 3,000 mn roubles at a rate of 537,000 roubles per kilometre. However, costs are falling as new pipelines are following existing routes along corridors where preliminary work (such as clearing) and research work have already been carried out. Costs will also be reduced with the introduction of year-round working which will enable better use to be made of pipelaying equipment. While Glavsibtruboprovodstroi built 9,780 kms of trunk and local pipelines in Western Siberia over 1976-1980, continuous working enabled it to lay 2,220 kms during the first half of 1981.

It is commonly asserted by Western journalists that permafrost creates problems in pipeline construction, but as one expert has pointed out, only 130 km out of the 53,000 planned for the last FYP period passed through zones of continuous permafrost.[1] While many pipelines pass through the discontinous permafrost zone, this does not pose such daunting construction problems.

---

1 W Sanders in Oil and Gas Journal 21/4/80.

Pipelines in continuous permafrost zones are either supported on wooden trestles or laid on a berm of sand. The problems of high winds are overcome by the installation of special vibration dampers on the trestles, but sand erosion on the berms is a problem. However it is considered to be merely an irritant and has not stopped the functioning of the pipelines.

In fact, the most difficult problem is swamps. Most of the Western Siberian lowland is swamp which until recently could only be crossed in winter when the ground froze over, and this reduced the construction period to five months a year. The arrival in Tyumen oblast of modern swamp vehicles is permitting year-round construction in most parts of the principal gas pipeline corridor Urengoi-Peregrebnoe. The other corridor, followed by the Urengoi-Chelyabinsk pipelines, runs alongside the Tyumen-Surgut-Urengoi railway, and seasonality of construction work has not been a problem.

In summer, gas pipelines laid in swamps tend to rise to the surface and split (this is not such a problem with oil pipelines because oil is heavier than gas) and they have to be weighted down by being cased in concrete. Normally up to 800 tons of concrete per km of pipe is required, although as much as 2,000 tons are occasionally needed. In some places, pipelines are fastened to the bottom of the swamp with harpoons. On the latest Urengoi-Gryazovets-Ivatsevichi line, an experiment has been carried out in which the pipeline is weighted by a polyethylene container which is wrapped round the pipe and filled with sand after the pipe has been trenched. The new method should reduce the time spent on the weighting process by 90%, and requires only 20 to 30 tons of polyethylene per km of pipeline compared with 800 tons of concrete. While it has proved adequate during short-term laboratory tests, long-term testing on a functioning pipeline is still continuing.

A more general problem is the low capacity of gas pipelines. While a 1,220 mm oil pipeline can carry 75 mn tons of oil a year when all its pumping stations are installed, a 1,420 mm gas pipeline can carry only 28 bn cu m of gas (23 mn tons of oil equivalent). This factor assumes crucial importance in view of the plans to raise Tyumen gas production from 140 bn cu m to 800 bn cu m a year over the next 20 years. To carry this gas to the western USSR, 24 pipelines will be needed, totalling 72-84,000 km. The minister of Minneftegazstroi, B. Shcherbina, has said that even if maximum use was made of compressor stations and the design pressure of 75 ats was maintained, allowing 35 bn cu m a year to be carried by each pipeline, 60,000 km of pipeline will still be needed.

This has concentrated Soviet minds on the need to raise the capacity of gas pipelines. While they have designed pipes of 1,620 and 2,520 mm (an example has been on display at Moscow's National Exhibition of Economic Achievements for many years) they have only recently mastered production of the 1,420 mm variety, which is currently the world's biggest mass-produced pipe. They have subsequently described any increase in pipeline diameter as "irrational", preferring to concentrate on other methods of raising throughput.

Three possible answers are the cooling of gas to minus 30 degrees (which reduces its volume), the raising of the pipeline pressure from the present 75 ats to pressures of up to 125 ats, and a reduction in the distance between compressor stations. A combination of these three possibilities would enable pipeline throughput to be trebled.

The construction of the first gas treatment installation incorporating a cooling system has been completed at Urengoi gas deposit. Its capacity is of 10 bn cu m a year, and it was intended to prepare gas for the fourth string of the Siyanie Severa pipeline to Moscow and the western border. The idea was to raise the pipeline's capacity and preserve the natural state of the tundra with the gas kept cool by a number of refrigerators installed on the pipeline. If the gas could be kept at 0 degrees, the pipeline's productivity could be raised by 7 to 10%, and the scheme envisaged the gas being kept at -1 to -3 degrees for a distance of 2,000 kms as far as Vologda. But the plan was not carried out. A possible reason was the failure to extend the 500 kv transmission line to the area from the Surgut power station, and without it the refrigerators could not be operated. The refrigeration plant at Urengoi may now be used to cool gas for the Yamal line to Western Europe.

Different types of multi-layer pipe have been developed. In Kiev, the Paton Institute invented a multi-layer pipe able to sustain pressures of 120 ats and carry gas cooled to -23 degrees, thereby doubling the pipeline's productivity. But this idea appears to have been shelved in favour of a 1,420 mm diameter pipe of 100 ats, able to carry 50 bn cu m a year.

These pipes are to be produced at the Vyksa plant, and Gosekspertiz has calculated that if all the pipelines to be built over 1981-1985 were to use this pipe, with compressor stations at optimal intervals of 90 to 100 kms, then the pipeline construction plan could be reduced by 33%, or by 15,000 kms.

An interesting variation has been developed by the Moscow-based Petrochemical and Gas Industry Research Institute; a double-walled pipe filled with concrete between the walls. It can withstand pressures of 155 ats and will enable 60 bn cu m of gas a year to be carried. It is particularly useful in the swampy conditions of Western Siberia, where the concrete prevents the pipe from rising to the surface in summer, and the laying of an experimental section has begun in North Tyumen. The new type of pipe also provides better insulation for cooled gas, but pipelines operating at such high pressures require quite different types of armatures, compressor stations, pumps and turbines compared with conventional pipelines.

In March 1981, a new generation of laminar pipe was tested on an experimental pipeline section of 1.6 kms near Kiev, with welding, insulation and trenching tests carried out. The Russians call it "quasi-monolithic" pipe manufactured during the pouring and rolling process with conventional technology. A batch has now been manufactured by the Khartsizsk plant for testing in northern Tyumen oblast.

The installation of more compressor stations of greater unit capacity on gas pipelines will play a crucial role in the achievement of gas production targets. The construction of compressor stations has traditionally lagged behind plans. Thus in 1979 it was planned to install 68 stations, but only 148 were built during the four years 1976-1979, and the plan for 1980 was set more realistically at 29. At the end of 1980 there were 494 compressors of total capacity 17,600 mw.

Soviet compressor stations have sets of 5,6, or 10 mw, but units of 16 mw are now being developed at the Sverdlovsk Turbomotornyi Zavod, and of 25 mw at the Nevskii Zavod in Leningrad. The testing of this last model took place for a period of two years at the Novgorod Experimental Compressor Station, and tests were said to be successfully completed in January 1982. Serial production is to begin later in the year, and the Russians have accelerated the tooling-up process at Nevskii Zavrod because they are afraid that the effects of American sanctions on European firms will delay the delivery of compressors for the Yamal pipeline. One of their options is to supply the necessary 25 mw units themselves, and take delivery of the foreign units for subsequent pipelines at a later date when American sanctions have been lifted.

A notable event of 1981 was the venture into compressor station construction by another Leningrad plant, LMZ, which began the production of 10 mw units for the Urengoi-Gryazovets-Moscow pipeline. The start up of the new facility will enable Soviet production of 10 mw units to double.

The "Turbogaz" plant of Uzhgorod also makes heavy duty compressors for gas pipe-lines, and one report states that it may become the main supplier for the Yamal pipeline.

Many large capacity compressor station units have been imported, although spare parts for them are made at the above-mentioned plants as well as specialised engineering plants like that at Chita which makes rotors for turbines. In 1980, the Russians spent $214.6 mn on the import of pumps for gas from OECD countries; France ($76.5 mn) was the main supplier, although British firms like John Brown and Rolls-Royce as well as German and Italian companies have sold equipment in the past.

The Russians are now showing great interest in aviation drive compressors because of their advantages over conventional models. A typical station of 50 mw capacity can be built in only five to six months compared with 16 to 19 months for the conventional type. They work on natural gas and are small and light, not needing foundations or buildings. Consequently they can be transported easily, and when a unit breaks down it can be quickly replaced from stock while it is being repaired. This avoids lengthy reductions in the capacity of the station which usually maintains a spare unit. The absence of water-cooling prevents freeze-ups, and the self-heating equipment enables it to work in the most difficult conditions. The first aviation drive units were installed in 1974, and by 1979 about 200 were employed at 32 stations, with a further 100 being installed at other stations.

Most of the USSR's aviation drive compressor stations have been imported from Western firms like Rolls Royce, which supplied the Urengoi-Chelyabinsk pipeline's stations. Soviet-made stations of 6 mv capacity employ engines taken directly from planes. However the "M.V. Frunze" Engineering Plant at Sumy in the Ukraine has begun the serial production of compressor units with capacities of 16 mw and able to pump 31 mn cu m a day. A report of December 1981 announced that about 200 Sumy units would be used for the Yamal pipeline, and this is presumably another Russian option in the event of Western firms being unable to fulfil their contractual obligations. Smaller compressors from Sumy are already oper-ating on the Nizhnyaya Tura-Perm-Gorki, Orenburg-Kuibyshev and Central Asia-Centre pipelines, but the new units will be three times more powerful. They are built in modular form, and can be installed extremely rapidly.

Aviation drive units have proved particularly useful for pipelines passing through inaccessible mountain areas. In 1979, they were installed at heights of 1,800 metres above sea-level on the pipeline carrying gas from the Stavropol fields to the Caucasus region. Installation took only five months. As well as for pipe-line compressor stations, they are also used in well-head pumping stations at aging deposits, and for pumping gas into underground reservoirs. Marine-type gas turbine motors provide similar advantages, and the Shipbuilding Ministry produced the first batch of 10 mw units for compressor stations in 1977.

Although the number of compressor units being built is increasing, they are unreliable, and their design requires improvements. Poor design is only one of a number of reasons for breakdowns, however, with labour and power shortages making themselves felt in Western Siberia.

At present, most compressor stations in Western Siberia serve up to three pipeline strings, and are staffed by the tour-of-duty method. This results in a constant turnover of labour. When the stations serve four or more pipelines (e.g. the Siyanie Severa), they need to employ 500-600 workers, and a permanent settlement has to be built. The construction of permanent settlements is the greatest single factor delaying the development of Western Siberia, and most compressor stations suffer from labour shortages.

The failure to provide North Tyumen with a permanent electricity supply has already been mentioned. It means that each compressor station must have its own mobile power station. Not only are these inadequate (a requirement of 600 mw capacity for a reliable power supply compares with an actual available capacity of 250 mw), but they also need a large number of personnel to maintain them. The result is constant stoppages at compressor stations. In the event of a power failure, a 6 mw unit can run for 15 minutes on batteries, but a 10 mw unit will last for not more than two minutes. Thus an unreliable power supply inhibits the installation of more powerful units. It takes 45 minutes to start them up again, and they can suffer stoppages up to five times a day. When it is remembered that they pump 1 mn cu m of gas an hour, the economic cost of such stoppages can be appreciated.

Power failures also damage the compressor units. As they grind to a halt, bearings become displaced, and the shaft decentralised. Some units have broken down within twelve hours, and the pipeline organisations have to employ an army of repair workers.

An answer to the power problem may be at hand with the inauguration of serial production at the Zvezda Plant, Leningrad, of a new type of large capacity mobile power station. Specially designed for northern Tyumen, it can be delivered by trailer or helicopter, and does not require foundations or special buildings. It is able to fully meet the power needs of a compressor station and settlement housing 2,000 people.

The rate of installation of compressor units should accelerate in the future as more use is being made of the modular assembly method. Large units of up to 600 tons are pre-assembled at the factory and then transported to the compressor station, where they are installed by the new specialised organisation, Sibkomplektmontazh. Labour costs are thus reduced by up to $2\frac{1}{2}$ times, the volume of construction materials by up to four times, and installation time by up to 35 per cent.

Control equipment at compressor stations is mainly made by the Soyuzgazavtomatika plant of Kaliningrad, which markets the "Impuls" and "Kvarts" systems.

It is hoped that in the future the distance between compressor stations can be reduced to 100 km, compared with the present average of 130 km, and, together with the installation of multi-layer pipe and the cooling of gas, this should enable appreciable increases in pipeline productivity to be achieved. In the meantime, efforts are being directed towards measures for reducing the unit cost of pipeline construction. These include the pre-insulation and improved cleaning of pipes, faster and more efficient welding, and better trenching equipment.

At the moment, most new pipelines are cleaned and insulated on site by special organisations belonging to Minneftegazstroi. Polyethylene and PVC tape are normally used. Only recently have pre-insulated pipes been received from the pipe plants, and pipe from Germany has sometimes been pre-insulated with hard polyethylene coatings. While this has often been damaged in transit, it can be repaired easily on site.

The cleaning process is said to be ineffective, often leaving sand, clay and ice in the pipelines. Dents and projections in the pipe prevent the use of mechanical cleaning methods. Water accumulates in the dips in the pipeline, reducing its effective capacity. On the Ust Balyk-Nizhnevartovsk oil pipeline, sand and water were removed by the use of highly viscous helium-based polymers, which served as highly successful cleaning and anti-friction agents.

Western experts have reported that Soviet pipelaying is by conventional practice, except for some aspects of welding. While American pipelayers employ double-jointing, i.e. the welding of two pipes together at a base plant, and then transporting the double-length pipe to the site of the pipeline, the Russians employ triple-jointing. This creates lengths of 36 metres which are hauled two and sometimes three at a time on long trailers pulled by tracked vehicles to the route of the pipeline. This means that 65 per cent of all welding can be carried out by automatic processes with only 35 per cent of joints welded by hand on site.

In the past, a shortage of pipe transporters has delayed pipeline construction. During 1980-1981, production of two new models began. The "Bakinskii Rabochii" plant of Baku has started making vehicles based on a Kroz lorry able to carry 3 pipes weighing up to 19 tons, and the Ukhtagazstroimash plant of Ukhta (a totally new plant) has begun to produce a machine specially designed to carry 1,420 mm pipe in Tyumen conditions. Based on a K-701 tractor, it is able to carry 3 pipes weighing over 100 tons.

The use of the new Sever-1 pipewelding complex is said to raise labour productivity five-fold and reduce pipewelding times considerably. It can weld 2.5 kms of pipe a day with stable high-quality welding operations in any weather, and it is claimed that it has not yet made a bad joint. It was invented by scientists from the Paton Institute in Kiev and tested on the first Urengoi-Chelyabinsk gas pipeline. Serial production of the machine began in 1978 and licenses have been sold to firms in many countries including the USA and Japan. Other plants have also designed semi-automatic welding machines for use in Tyumen oblast, including the Vilnius Electric Arc-Welding Equipment Plant (which makes 40% of the arc welding machines produced in the USSR) and the Pskov Heavy Electric Welding Plant.

Raising the efficiency of other aspects of pipeline construction work is held back by the fact that Minneftegazstroi produces only one-off and small-batch machines, and the serial production of equipment is undertaken by engineering ministries who do not always fulfill their obligations. There are constant complaints in the Soviet press about armatures, of which large numbers are required (they are fitted every 30 km on average), and which usually weigh 50 tons. They need to be much lighter for installation work to be speeded up, and their range needs to be extended because they are not produced for some types of pipe, especially those with large diameters.

While most traditional models are still made at various plants in Baku and the Georgiyevsk Armature Plant in Stavroipol krai, new plants are being built at Pugachevo (Saratov oblast) where the first stage has been commissioned of what will become the USSR's largest valve plant, Zaporozhe where the existing factory is being extended to produce heavy duty valves for large diameter pipelines, and Ust Kamenogorsk, where the first stage with an annual output of 225,000 valves a year has started up. The USSR also imports valves from firms like Borsig of West Germany.

More rotary excavators capable of shifting 1,200 cu m of earth an hour are needed for trenching, and cranes and tracked vehicles of greater capacity are required. A powerful excavator of new design, intended for permafrost regions and built by the Moscow experimental plant of Minneftegazstroi, was used for the second string of the Urengoi-Chelyabinsk pipeline. Called the EF-131, it was created on the basis of the T-130 tractor and can dig up to 3 km of trench a day. Serial production must begin soon if gasworkers are to tap the Far North deposits during the Eleventh FYP period. Meanwhile, the same plant has also designed a new rotary excavator, the ETR-254, which can move 1,200 cu m an hour of frozen ground up to 3 metres deep. Production has begun at a rate of 44 a year.

The other significant producer of excavators is the "Talleks" enterprise of Tallin. It manufactures 200 machines a year which are faster than those from Moscow, but less powerful, digging down to only 2 metres. While conventional excavators have been used for pipeline repair work in the past, the Bryansk Plant of Irrigation Machinery has started making a specialised machine with two rotors. Designed by the |Vnizemmash| Institute of Leningrad, it can move 425 cu m an hour of earth, and has been tested on the Druzhba oil pipeline.

A waterpumping unit, intended to assist pipeline construction in swamps, has been designed by another engineering enterprise in Bryansk. Mounted on a DT-75 tractor, it can pump 700 cu m of water an hour, and the first batch was employed on the Urengoi-Gryazovets-Moscow pipeline. The "Gazstroimashin" Special Design Bureau of Minneftegazstroi is currently working on improving the machine, particularly by increasing its engine power.

In 1981, the year-round construction of pipelines in Tyumen oblast was attempted, and this has raised the need for general-purpose swamp vehicles. Hopes rest mainly on the "Tyumen", built by Minneftegazstroi's Kropotkin Engineering Plant in the North Caucasus. Designed by the Gazstroimashin Bureau, it was extensively tested during the laying of the Surgut-Polotsk oil pipeline. It is based on the Kirovets tractor and has twin bogies with autonomous drive; if one sinks in the swamp, the other pulls it out, and its track load of only 300 grams per square cm should enable it to travel most places. It has a carrying capacity of 36 tons and is able to transport stacks of drill pipes, excavators, welding machines etc. at a maximum speed of 15 kms/hr. The first batch of vehicles arrived in Tyumen oblast in May 1981, and the Kropotkin Plant aims to expand its current production capacity of 100 a year. Similar vehicles are to be made by the "Oil and Transport Engineering Plant" of Tyumen, the construction of which began in 1981. When completed, it will also make dredgers and pipe carriers.

Sterlitamak construction machinery plant has begun serial production of pipelaying machines able to carry more than 50 tons. For more than three years, prototypes were tested on the construction of Siberian oil and gas pipelines, and they are said to have proven highly successful, even in temperatures ranging from -50 to +40 degrees.

Serial output of these machines, based on the T-330 tractor produced by the Cheboksary Heavy Tractor Plant, began in 1980 when 26 were produced. In 1981, 115 were turned out. With the commissioning of the plant's second stage at the end of 1981, it is expected that 230 machines will be manufactured in 1982, and the Eleventh FYP has called for output to rise to 500 a year by 1985.

The Bryansk Tractor Plant makes pipelayers based on its T-130 tractor. They are designed to lay pipes of 1,420 mm diameter, and the first machines have been sent to Tyumen oblast. In 1982, it plans to build 500 pipelayers. The Baku Experimental Plant has concentrated on designing improved machines for laying lighter pipe. Its latest model has a carrying capacity of 15 tons, and can lay pipe of up to 720 mm diameter faster and cheaper than previous models.

Perhaps the most important development in pipeline technology is the invention of a floating pipelaying complex that can lay pipelines all year round across the Siberian swamps. The complex consists of four vessels. The first carries an excavator which cuts a shipping channel. The second is a floating crane, and the third is a platform on which the crane places pipe delivered by helicopter. The pipes are welded and insulated on the platform. The fourth stage is a pontoon carrying the completed pipeline. Every time a new section is welded, the pontoon is pulled forward and the pipeline slips into the marsh. The source does not say whether the pipeline is then covered; it may in fact be weighted with concrete so that it stays at the bottom of the shipping channel. The new system has laid pipes at a rate of 1 km a day during tests near Tyumen.

The degree of plan fulfillment by suppliers must be improved; while some plants, like the Kolushchenko engineering plant, Chelyabinsk, fulfill their obligations for the delivery of bulldozers, others, like the Uralsk auto plant, have fallen behind on the delivery of vehicles. The organisation of repairs to equipment must be radically improved.

But in spite of scope of improvement, it can be said that Soviet pipelaying work is fast and efficient. In terms of technological comparisons, one Western expert has said, "Arctic or winter pipeline construction procedures in the Soviet Union and North America do not materially differ. It can be safely said that they don't know anything we don't know and vice versa."[1]

Much of the above also applies to oil pipelines. The growing use of modular units has enabled pumping stations to be installed at accelerating rates: in 1976 it was planned to build 16, and in 1977 30 stations were completed, including 15 on the Nizhnevartovsk-Kuibyshev pipeline. The plan for 1978 was to install 31 including the remaining nine on the Nizhnevartovsk-Kuibyshev line, and the plan for 1979 amounted to 36. In fact, there has been a significant degree of plant underfulfillment, with only 100 stations coming on-stream during the four years 1976-1979, or 30 less than planned.

The size of pumping stations is increasing. The standard station is now equipped with centrifugal pumps of up to 12,500 cu m an hour capacity and 8 mw electrical motors working at 3,000 revs/minute. The "Elektromash" plant of Tiraspol has been reconstructed to enable the production of large electric motors for pumping stations.

The USSR also imports pumping stations, with eight from Vagonka-Poprad plant of Trevisov, Czechoslovakia, now operating on Siberian pipelines. A further 45 will be imported during 1981-1985.

A new type of open pumping unit (i.e. without protective building) which can operate in temperatures of -60 degrees has recently begun to be installed. It was success-fully tested at the head of the Usa-Ukhta-Yaroslavi pipeline. It is mounted on an open platform and is heated by its own motors with heat which was formerly wasted. A whole pumping station can now be erected in seven months instead of the usual 37.

The automation of pumping stations is being accomplished with the "Pusk-71" system which can be used for both head and pipeline stations with any number of pumping units. During the period 1976-1980, it was installed at 160 stations.

OIL PIPELINES

Oil pipelines laid during the Tenth and Eleventh FYP periods have consisted almost entirely of large diameter pipelines carrying oil out of Western Siberia, and those redistributing this oil among the refineries in the heavily populated European part of the USSR so as to achieve a better distribution of oil product output.

Oil production in Western Siberia has been rising by 30 mn tons a year and none of this oil can be carried by rail because the only railway out of the oil-producing

---

1 Sanders op. cit.

area is working at capacity. Therefore it all has to be moved by pipeline, and these must be built at a rate of one 1,220 mm pipeline every two years. Policy has been to strengthen the existing Western Siberia-Volga network by the construction of two strings from Nizhnevartovsk to Kuibyshev.

These pipelines are not particularly long by Soviet standards. They run for 2,089 km along the Nizhnevartovsk-Surgut pipeline corridor, and then alongside the Tyumen-Surgut railway as far as Yarkovo, at which point they continue in a south-westerly direction to Kurgan. From here, they run almost due west through Chelyabinsk and Ufa to Kuibyshev, a major oil-refining and industrial city on the Volga.

The first pipeline was completed in 1976, and by November 1978 the second string was working at its planned capacity with all pumping stations in operation. These two pipelines have a combined capacity of 150 mn tons a year, and are transporting most of the increase in oil production from Western Siberia during the last FYP period.

The redistribution of oil from these two pipelines has been influenced by the regional distribution of new refinery construction, which in turn reflects the need to optimise the regional production of oil products.

Formerly, the Ukraine was badly served by refineries, with a number of smaller ones operating on local oil from declining fields. This has been remedied during the last five years, with refineries at Lisichansk and Kremenchug being built, and those at Odessa and Kherson being extended. Accordingly, a large diameter pipeline has been built from Kuibyshev down through Central Russia and the Ukraine to Odessa.

The line was built in three sections. The first section, running for 950 km from Kuibyshev to the Lisichansk refinery, was completed in July 1977 and the second (Lisichansk to Kremenchug) was ready by June 1978. The third section runs from Kremenchug to Kherson, and at the Snigirovka pumping station a 200 km branch line begins its journey to Odessa. The whole system, from Samotlor to Odessa, is 4,100 km long and also provides oil for export from the oil port of Odessa.

As well as optimising refinery location, the USSR faced the need to keep existing refineries working at full capacity in regions with declining oil production. This has led to the decision to direct the second major distribution pipeline from Kuibyshev down towards the important refineries of the North Caucasus, and on to the oil export base of Novorossiisk. From here it is being exported to Mediterranean countries, including Italy and France.

The Kuibyshev-Tikhoretsk-Novorossiisk pipeline, nearly 1,500 kms in length and 820 mm in diameter, was completed in 1974, and the last section of the Rostov-Groznyi pipeline, which crosses it at Tikhoretsk, was finished in 1981. It is now possible for Siberian oil to flow to the Groznyi refineries, and this explains why they are being expanded despite the decline in local oil production. The pipeline is now being extended south-eastwards to Baku, where the Baku 22nd Congress refinery was given a new first-phase unit in 1981.

The latest pipeline out of Western Siberia has followed a quite different route, from Surgut to Perm and then on to Polotsk near the western border of the USSR. The pipeline stretches for 3,300 kms with a diameter of 1,220 mm and it serves the Perm, Gorki, Yaroslavl, Moscow, Ryazan and Novopolotsk refineries. It also provides oil for the Kirishi and Mazeikiai refineries as well as for export through the Ventspils terminal.

The pipeline was begun in September 1977. The first section of 1,256 kms from Surgut to Perm was completed in May 1979, and oil reached the Perm refinery through a 24 km branch line in August 1979. The next section, of 818 km to Gorki, was completed in February 1980, and the pipeline's throughput could then be increased considerably because oil could now flow to Yaroslavl and on to Kirishi, near Leningrad, through an existing pipeline. Kirishi's capacity had been increased by 6 mn tons a year to 18 mn tons a year in June 1980 so as to take Siberian oil.

Work on the third section from Gorki to Polotsk began in November 1979. It was carried out by the Mosgazprovodstroi and Lengazspetsstroi organisations from Moscow and Leningrad, helped by workers of the Polish enterprise "Energopol" who were responsible for building the 300 km stretch from Andreapol to Novopolotsk. The Poles had previously built the 442 km pipeline from Novopolotsk to Mazheikiai, and Poland will get 1 mn tons a year as payment for its contribution.

The laying of the Surgut-Polotsk line was completed in February 1981, and the first oil reached Novopolotsk the following April. It will eventually have 32 pumping stations, a reservoir capacity of 1 mn cu m and 2,000 km of electricity transmission lines, and will carry 75 mn tons of oil a year.

In view of the fact that building pipelines in multi-string corridors leads to considerable reductions in unit costs, it can be surmised that Surgut-Polotsk is just the first of a group, maybe up to four, following this route. Hints of a second have already appeared; Georgian specialists have designed a new oil pipeline 3,500 km long, which is intended to cross the USSR and Poland, and which will be built by Polish and Russian workers, and the Eleventh FYP also refers to another pipeline from Kholmogorsk in Siberia to the Mid-Volga region, following the Surgut-Perm corridor.

Other important oil pipelines completed during the Tenth FYP include that from Omsk to Pavlodar refinery in northern Kazakhstan. 450 km long, it was completed in September 1977, with Samotlor oil arriving at Pavlodar in October of that year. As long ago as 1976, work began on building a pipeline from the Chardzhou refinery in the Turkmen republic (now under construction) through Chimkent, where another refinery is to be built, to Pavlodar. But the building of the refineries was postponed and the laying of the pipeline has accordingly been put back to 1983.

Another string of the 1,400 kms Uzen-Gurev-Kuibyshev pipeline was completed in 1981. This pipeline proved necessary because the first string frequently sprang leaks due to the conventional insulation being unable to withstand the high temperature of the oil. The oil is hot because 19 boilers have to be fitted to the pipeline at intervals of 50 to 100 kms to keep the oil, which has a high paraffin content, from congealing.

Important pipelines of local significance built during the period include that from Kalamkas on the Buzachi Peninsular to Shevchenko, which was finished in 1979. A 180 km pipeline from Kiengop to Naberezhnyi Chelny will give more flexibility in rearranging oil movements in the European USSR in connection with the increasing inflow of oil from Siberia. A 400 km pipeline from the Samgori oilfield in Georgia to Batumi was finished in late 1980, enabling part of Georgia's growing output to be refined and part to be exported.

A 75 km underwater pipeline was built to connect the Neftyanye Kamny collection centre in the Caspian Sea to the Baku refineries, and an 84 km line was built between the Bulla Island deposit and the Azerbaidzhan shore. They brought the aggregate length of underwater pipelines in the Caspian to more than 1,000 kms and will allow a fleet of tankers to be used elsewhere, cut oil losses during transportation, and permit a more regular flow of oil from the oilfields to the refineries.

The recent sharp fall in output by Perm oblast may be reversed with the completion of a line carrying oil from the new Berezniki field to Perm refinery. In Western Siberia, the development of the new Krasnoleninsk field has been helped by the completion of a line carrying the oil to the Shaim field, where it will enter an existing line to Tyumen. Until recently, Shaim oil has been transferred to the railway at Tyumen, but at the end of 1981, a new pipeline connecting Tyumen with the Trans-Siberian line at Yurgamysh near Kurgan was completed. Oil from Krasnoleninsk can now flow directly to the Volga refineries.

There appear to have been few developments in extending the oil product pipeline network. At the moment, it consists of two major pipelines (Ufa-Omsk-Novosibirsk, 3,000 km long and 529 mm in diameter, and Kuibyshev-Bryansk, 1,137 km long and 529 mm diameter) and a number of shorter ones. Soviet policy has been to end the transportation of crude oil by the railways before starting the more difficult task of reducing the volume of oil products carried by rail. However, the Eleventh FYP foresees the construction of 9,000 kms of new product lines. They include a 400 km line from Pavlodar refinery to Semipalatinsk to supply farms in eastern Kazakhstan with petrol and diesel, and a 228 km line from Gorki refinery to Vladimir. This line was finished in September 1981: it has a capacity of 1.5 mn tons a year and has allowed 800 rail cisterns to be used elsewhere.

The pipeline from Yuzhnyi Balyk to Tobolsk petrochemical plant was nearing completion in April 1980. It will carry non-stable benzine from the Western Siberian gas refineries to Tobolsk; this is currently railed from Yuzhnyi Balyk to petrochemical plants in the Volga region. As the Tobolsk plant will not start up for some time, the benzine will be transshipped to the railway at Tobolsk, and the pipeline will thus enable severe congestion on the Surgut-Tobolsk section of the railway to be relieved.

GAS PIPELINES

As with oil pipelines, most new gas pipeline systems are designed to carry gas out of Western Siberia and distribute it throughout the European part of the Soviet Union. There are three such systems: Siyanie Severa, Tyumen-Moscow, and Tyumen-South Urals.

The Siyanie Severa (Northern Lights) system originally carried gas from the Vuktyl deposit in the Komi republic down to Torzhok, where some of the gas was diverted to Leningrad and Moscow, and then on to Minsk, Ivatsevichi in the Western Ukraine, and Dolina on the Czech border, from whence it was exported. During the last few years, the system has been expanded, but has continued to follow this route. It has been extended eastwards, firstly to the Medvezhe gas deposit in 1972 by the USSR's first 1,420 mm diameter pipeline, and then to Urengoi in 1977.

The most recent project is Siyanie Severa-4 which was begun at Urengoi in November 1979, and reached Vuktyl in July 1980. It is currently being built at a rate of 4 kms a day, and will follow the whole length of the route down to the Czech border, i.e. 4,400 kms. It will be completed, together with its 30 compressor stations, when the final 1,330 km section from Gryazovets to Ivatsevichi comes on-stream in 1982.

Torzhok has been succeeded by Gryazovets as the point at which gas is syphoned off for Leningrad and Moscow. A 620 km pipeline, capacity 9 bn cu m a year, from Gryazovets to Leningrad was completed in 1979, and an important branch of

Siyanie Severa-4 is the Gryazovets-Moscow pipeline of 457 km which was completed in February 1981. It carries 14 bn cu m of gas a year for distribution throughout Moscow oblast by a new 158 km Moscow Ring Pipeline. In summer, some of the gas is stored in Shchelkovo gas reservoir to be drawn on in winter.

In 1969, when Siyanie Severa began operating, its capacity was 2.9 bn cu m a year. By 1981, it was 75 bn cu m a year, and will rise to 91 bn when Siyanie Severa-4 is completed in 1982.

The second system (i.e. Tyumen-Moscow), branches from Siyanie Severa at the Peregrebnoe compressor station on the River Ob, and travels south to Nizhnyaya Tura where it splits, with two pipelines continuing their southerly course through Sverdlovsk to Chelyabinsk, and the other two following a westerly path through Perm, Kazan and Gorki to Moscow. The second Perm-Kazan-Gorki pipeline was completed in 1979, and most of the new lines to be built during the Eleventh FYP period will follow a corridor which can be considered a variant of the Tyumen-Moscow system. From Nizhnyaya Tura the pipelines follow a new course to join the Tyumen-South Urals system at Dyurtyuli.

The Tyumen-South Urals system runs due south for 1,600 km from Urengoi to Chelyabinsk and then on through Petrovsk to the eastern Ukrainian town of Novopskov where it meets the Soyuz pipeline.

It initially ran from the medium-sized gas deposit of Vyngapur, which was opened up while Urengoi was being prepared. The Vyngapur-Chelyabinsk pipeline was begun in February 1977, and was built very rapidly because it follows the Tyumen-Surgut-Urengoi railway line for much of the way. It reached Chelyabinsk in November 1978 by which time work had begun on three projects: to extend it for 380 km northwards to Urengoi; to build a second string in the same way, i.e. from Vyngapur to Chelyabinsk and then Urengoi to Vyngapur (this was finished in June 1980); and to extend the first string southwards to Petrovsk (completed February 1980) and further to Novopskov which it reached in August 1980. The pipeline carries 30 bn cu m/year.

The throughput of the Tyumen-South Urals system was expected to grow from 21 bn cu m in 1979 to 52 bn cu m/year in 1980. One of its aims is to enable Siberian gas to make up for the declining production of the Ukrainian gas fields, and also to reverse the flow in the North Caucasus-Moscow pipelines, so as to feed Siberian gas to the Caucasian republics. This is proving necessary because of the halt in imports from Iran.

The Soyuz pipeline, running 3,300 km from Orenburg in the South Urals to the western frontier at Uzhgorod was, like all the above-mentioned pipelines, built of 1,420 mm pipe. It was completed in December 1978. Of its projected capacity of 28 bn cu m/year, 15.5 bn cu m/year are to be sent to the Eastern European countries which participated in its construction. East Germany, Poland, Hungary, Czechoslovakia and Bulgaria get 2.8 bn cu m/year each for having undertaken the complete construction of sections of the pipeline, and Rumania gets 1.5 bn cu m for having supplied equipment.

Another string of the Central Asia-Centre pipeline has been built during the period. As well as these major trunk pipelines, a number of small but vitally important projects have been completed.

Branch lines from the three major trunk systems from Western Siberia are being built to every major town along the routes. One line which has attracted much attention is that of Orsha on the Siyanie Severa to the rapidly growing Belorussian town of Vitebsk. It was completed in summer 1980 and permitted a reduction in the use of low-grade fuels like peat and firewood. Other important branch lines run from the Siyanie Severa corridor for 90 kms to the medium-sized Belorussian town of Mogilev, which has a large synthetic fibres plant, and from the Tyumen-Moscow system for 383 kms to the important regional centre of Kirov with a population of over 400,000.

The Shurtan-Syr Darya pipeline in Uzbekistan, completed in October 1980, is important because of the shortage of fuel oil in Central Asia. It carries 6.2 bn cu m of gas a year from gasfields in the Karsh Steppe over 400 kms to Tashkent's largest power station, the Syr Darya, which has been running at half capacity because of the shortage of fuel oil, and has since been converted to natural gas. The pipeline is to be extended to Tashkent and will supply 2 bn cu m a year to Taskhent Power Station. A branch to Leninabad in the Tadzhik republic will be commissioned shortly.

At the gas pipeline connecting the world's northernmost deposits of Solenoe and Messoyakh with the rapidly growing copper and nickel mining town of Norilsk, a third string has been built. This pipeline lies entirely in the continuous permafrost zone and rests above ground on wooden trestles. Construction of a fourth string has now started.

The Mozdok-Kazi Magomed pipeline, with a length of 680 kms and a diameter of 1,220 mm was started in late 1980. It will carry 13 bn cu m a year of Siberian and Orenburg gas from Novopskov through the North Caucasus-Moscow line (on which the direction of flow has been reversed), via Rostov, Stavropol and Mozdok to the Caucasus republics, and this will more than make up for the loss of 10 bn cu m a year, formerly supplied by Iran. It is unlikely that Iranian supplies will be resumed for several years as Iran's oil industry (from which the gas is obtained) has all but collapsed. There is no hope that the Consortium Pipeline project, under which Iran would send 17.2 bn cu m/year to the USSR through the Igat-3 pipeline with the USSR sending a similar amount to Western Europe through Czechoslovakia, can now be carried out.

The Yelvakh-Nakhichevan pipeline runs for 200 km in the Caucasus and has had to overcome many problems connected with the mountainous terrain; its highest point is 1,934 metres above sea level. It was commissioned in December 1980.

While the three Siberian trunk systems shoulder the burder of supplying European Russia, the Ukraine is being allowed to cover only its own requirements. To this end, a pipeline has been laid for 600 km from the Shebelinka deposit in a south-westerly direction through Dnepropetrovsk and Krivoi Rog to Ismail. It was completed in 1979. The Ukraine network has been extended to the Crimean resort of Yalta, with the aim of improving the air.

The Nizhnevartovsk-Kuzbas pipeline is important in two respects. Firstly, it enables the vast reserves of casinghead gas produced at the Western Siberian oilfields to be used, whereas formerly they were flared. And, secondly, it brings piped gas for the first time to the large industrial area of the Kuzbas. But the oilfields are not yet producing sufficient oil well gas for the pipeline to be operated at capacity, and it was only in 1980 that the first compressor station on the 1,400 pipeline was opened at Parabel. It is expected that this will enable throughput to be doubled. By August 1980, a 150 km branch line reached Novosibirsk, which will receive 1 bn cu m of gas in 1981 and more than 2 bn cu m in 1982.

Construction of a 340 km pipeline from Vilnius to Kaliningrad is continuing. Built by Lengazspetsstroi, it reached Kaunus in early 1981, and should deliver Siberian gas to Kalinigrad by mid-1983.

PIPELINE PLANS

According to the Eleventh FYP, 10,000 kms of crude pipelines are to be built during 1981-1985. This includes the 245 km Tyumen-Yurgamysh line (now completed), the remaining sections of the Groznyi-Baku line, and a new line from Kholmogorsk through Surgut and Perm to Almetyevsk. The Omsk-Pavlodar line will be extended for 2,300 kms to Chimkent and Chardzhou, and another pipeline from the Tyumen oilfields to Central Russia will be built. A further 9,000 kms of product pipelines are planned, mostly short lines between refineries and major industrial areas, particularly where existing rail links are congested. A significant share of the total will consist of product lines serving farms, particularly in areas where the road network is not good. During the five year period, 97 more pumping stations and 6.7 mn cu m of new storage capacity are planned.

The original FYP target of 49,500 kms of new gas pipelines has been scaled down to 45,000 kms. More than half (26,000 kms) of this will consist of 7 trunk lines, each of 1,420 mm diameter, running broadly westwards from Urengoi. The first of these was Siyanie Severa-4, which will be completed in the summer of 1982, and the second was Urengoi-Petrovsk (2,740 kms) which was finished in December 1981. It required only 410 days to build at a rate of 6.68 kms a day, and the line was laid 6 months ahead of schedule. Construction of the compressor stations is also said to be proceeding ahead of schedule, and when completed, they will pump 32 bn cu m a year to Central Russia.

The third of the seven pipelines will follow the new Urengoi-Petrovsk corridor, but will continue to Novopskov, and will enable Siberian gas flow to the eastern Ukraine to be increased. Work began in January 1982 and got off to a good start with the first 300 kms trenched by the end of the month.

The fourth pipeline is known as the "Yamal Pipeline" in the West (the Russians call it "Urengoi-Uzhgorod", or the "Export Pipeline") and will run for 4,120 kms to the western border of the USSR, and then across Czechoslovakia to Western Europe. It will follow the "Tyumen-Moscow" corridor as far as Kazan, and then travel south-westwards through Yelets and Kursk to Kiev, where it will join an existing system of pipelines to Uzhgorod. The first kilometre of pipe was trenched on 12th February, and the whole line is to be laid by the autumn of 1983, with all 41 compressor stations working by October 1984. Pipelaying will take place simultaneously on 10 sections, with pipe and equipment delivered to the line at 51 field sites. A new association, Soyuzintertransgaz, has been set up to administer construction work, and supply bases are being built at Nadym in Western Siberia, Pomary in the Mari republic, Torbeyevo in the Mordovian republic, Kursk and Kiev.

Two more pipelines will follow the Urengoi-Petrovsk line, before striking due west to the distribution centre at Yelets, and the 7th trunk line will be the first to start from the supergiant Yamburg field in Western Siberia. It will follow the path of the Yamal pipeline, and is designed to carry gas exports to Eastern Europe.

During 1981-1985, it is planned to build 350 compressor stations of total capacity 20,000 mw, and 227 of these (capacity more than 12,000 mw) will be installed on the seven trunk lines to be built out of Urengoi. In 1981, stations with a capacity of 2,400 mw were completed (taking total capacity to 20,000 mw, or 13.6 per cent more than at the end of 1980), and the 1982 plan expects a further 4,200 mw to be added. This implies that an average of 4,500 mw a year must be built over 1983-1985 for the FYP target to be met. If the manufacture and commissioning of the Nevskii Plant's new 25 mw model goes smoothly, then this target should prove well within reach.

CHAPTER 4

The Refinery Sector

LOCATION OF OIL REFINERIES

The USSR does not issue statistics on oil refining, but it can be estimated that their 44 operating refineries processed 495 mn tons of oil in 1981 and had a total capacity of 561 mn tons at year end.

During the five years 1976-1980, the output of oil products grew by 3.7 per cent a year compared with a growth in crude oil production of 4.2 per cent a year. The Eleventh FYP for 1981-1985 calls for an increase of 2.0 per cent a year in product output, and given the planned 1.0 per cent a year increase in the output of crude oil, it would appear that the Russians expect to reduce their exports of crude in favour of higher product exports.

The location of the oil refining sector is still heavily concentrated on the traditional refining regions of the Volga, Azerbaidzhan and the North Caucasus. Western Siberia, which produces half the country's oil, has only one refinery, at Omsk. This heavy concentration of refining capacity in a limited number of regions puts a heavy strain on the railway system for the transport of oil products. To supply Tashkent, for example, oil products must be hauled 2,000 km from Pavlodar or Krasnovodsk.

With the railway system under increasing strain, there has been a slight change in Soviet refinery location policy during the 1970s. While the former policy of increasing total capacity by expanding existing plant is being continued, more and more emphasis is being placed on the construction of new refineries. Four have come on-stream during the Tenth FYP period; these are Lisichansk (Ukraine) in 1976 and doubled up to 12 mn tons in 1979, Mozyr (Belorussia) in 1976. Pavlodar (Kazakhstan) in 1978 and Mazeikiai (Lithuania) in 1980. All are equipped with the standard 6 mn ton capacity atmospheric vacuum primary distillation plant.

Another refinery at Achinsk (Eastern Siberia) was due to start up in 1981 but has been deferred to 1982, and plans for others at Chimkent (Kazakhstan) and Chardzhou (Central Asia) have been greatly delayed. The assembly of equipment began at Chardzhou in March 1980 with East German experts participating, and the refinery is scheduled for start up in 1984. It is actually located at the new town of Neftezavodsk, 90 kms from Chardzhou. The Chimkent refinery should be commissioned in 1983 according to current plans.

There have been reports in the past that refineries are to be built at Arkhangelsk, Nakhodka, Vinnitsa and Zaporozhye, but no construction work appears to have been carried out yet.

Table 16

Oil Refineries in Operation, 1980

(mn tons)

Refinery	Region	Estimated capacity
Kirishi	Leningrad	18
Ukhta	Komi	6
Mazeikiai	Lithuania	6
Novopolotsk	Belorussia	20
Mozyr	Belorussia	12
Moscow	Central	12
Ryazan	Central	15
Yaroslavl	Central	24
Groznyi Sheripov		
Novo-Groznyi Lenin	North Caucasus	24
Tuapse	North Caucasus	6
Baku 22nd Congress	Azerbaidzhan	26
Novo-Baku Vladimir		
Ilyich	Azerbaidzhan	18
Batumi	Georgia	6
Tbilisi	Georgia	3
Novo-Gorki	Volga Vyatka	24
Kremenchug	Ukraine	15
Kherson	Ukraine	9
Lisichansk	Ukraine	18
Odessa	Ukraine	6
Drogobych	Ukraine	3
Lvov & Nadvornaya	Ukraine	6
Perm	Urals	24
Krasnokamsk	Urals	6
Orsk	Urals	12
Ufa	Volga	
Ufa 22nd Congress	Volga	46
Novo-Ufa Lenin	Volga	
Saratov	Volga	15
Volgograd	Volga	12
Novo-Kuibyshev	Volga	30
Kuibyshev	Volga	6
Syzran	Volga	6
Ishimbai-Salavat	Volga	12
Nizhnekamsk	Volga	12
Omsk	Western Siberia	30
Angarsk	Eastern Siberia	30
Komsomolsk	Far East	3
Khabarovsk	Far East	9
Gurev	Kazakhstan	3
Pavlodar	Kazakhstan	12
Krasnovodsk	Central Asia	12
Fergana	Central Asia	4
Total		561

The 44 refineries are:

Kirishi (Leningrad oblast)

Situated 114 kms from Leningrad, Kirishi started up in 1966, and receives oil by pipeline from Yaroslavl. During the tenth FYP period, a para-xylene plant (the first of its kind in the USSR), an octafiner and a hydrogen plant were completed by Kawasaki in 1976, a gas liquefaction and bottling plant was added in 1977, and in 1980 a new first-phase plant of capacity 6 mn tons a year and a parex plant for producing 120,000 tons of liquid paraffin a year were installed. In 1981, a catformer of 1.05 mn tons a year was added. The capacity of the refinery is now 18 mn tons a year.

Ukhta (Komi republic)

This is a small refinery which dates from 1939 and processes local oil and gas condensate into 8 different products. It has a capacity of 6 mn tons a year, and most oil from the Komi oilfield is sent to Yaroslavl. A large parex plant is under construction; its liquid paraffin will supply a new synthetic detergent plant under construction at Ukhta.

Mazheikiai (Lithuania)

Construction began in 1970, and after lengthy delays the refinery finally began working in May 1980 when fuel oil was delivered to the nearby Akmyants cement works. The first stage, producing petrol, diesel, kerosene, lubricants and LPG was completed in August 1980. The second stage is now under construction and will begin producing shortly. The refinery currently satisfies Lithuania's petrol and diesel requirements and exports fuel oil and LPG. It has a throughput of 6 mn tons a year, and when this is built up to the rated 24 mn tons a year, it will become the Soviet Union's largest source of products for export.

Novopolotsk (Belorussia)

Novopolotsk started up in 1963 on oil arriving through a branch of the Druzhba pipeline. Its size has increased considerably during the last five years in preparation for the completion of the Surgut-Polotsk pipeline now under construction. The capacity of the Druzhba pipeline has also been increased, and the refinery can now handle 20 million tons of oil a year. It is technologically advanced, with a large volume of secondary refining capacity, and produces chemical feedstocks for the rapidly growing Belorussian petro-chemical industry, as well as 120,000 tons a year of liquid paraffin from a parex plant.

Mozyr (Belorussia)

The Mozyr refinery began operating in 1975 and was supplied by a pipeline bringing oil from the nearby Rechitsa field. Local oil output has been declining, however, and the refinery is now supplied mostly from the Druzhba pipeline. In 1976, the construction of the second stage began, and its main units came on-stream in 1979, bringing the refinery's capacity to 12 m.t. Mozyr produces large volumes of petro-chemicals, has a 500,000 ton a year bitumen plant and produces raw materials for animal foodstuffs from its parex plant.

Moscow

This is an old refinery, built in 1934. Its capacity was doubled in 1976 in preparation for oil coming from the Komi republic through the Ukhta-Yaroslavl-Moscow pipeline. It has recently begun to receive Tyumen oil from the Surgut-Polotsk pipeline, and its current capacity is 12 mn tons a year. The Russians are upgrading the plant, and a large volume of secondary processing equipment will be installed during the 1980's. A vacuum distillation plant and a 1 mn ton a year catcracker are now being built.

Ryazan (Central Region)

This medium-sized refinery of 15 m.t. a year capacity consists basically of primary processing plant. It uses Volga oil and will be supplied from Tyumen shortly when a branch line from the Surgut-Polotsk pipeline is completed. A new shop for the production of lubricants came on-stream in 1980.

Yaroslavl (Central Region)

The Yaroslavl refinery works on oil from the Volga region, the Komi republic and Tyumen oblast. Oil has recently started arriving through the Surgut-Polotsk pipeline. The refinery produces petro-chemical feedstocks for Yaroslavl's synthetic rubber plant, and has a primary refining capacity of 24 mn tons a year. It has 13 facilities including large catcrackers and catformers producing light and medium products for most of the Central region.

Groznyi Sheripov (North Caucasus)

One of the oldest refineries in the USSR, dating from the last century, Groznyi Sheripov was reconstructed in 1974, and its annual output was increased by 250%. A large volume of secondary refining capacity was installed, so as to enable increasing quantities of light products to be obtained from a declining volume of crude.

Novo-Groznyi Lenin (North Caucasus)

Novo-Groznyi Lenin has also been undergoing an extensive reconstruction, and in 1979 a new first-phase facility was added, bringing the total capacity of the Groznyi plants to 24 m.t.

Tuapse (Krasnodar, North Caucasus)

A small plant of capacity 6 m.t. a year.

Baku 22nd Congress (Azerbaidzhan)

The refinery now has a capacity of 26 mn tons a year with the start up of a new first-phase unit with electric desalter in July 1981. A catformer of 1.05 mn tons a year, a vacuum distillation unit for producing lubes (which have been in short supply in the Caucasus region) and a coking chamber for the production of petroleum coke were also commissioned in 1981.

Novo-Baku Vladimir Ilyich (Azerbaidzhan)

The 1972 Girproazneft plan for the reconstruction of the Baku refining industry called for the capacity of this refinery to be doubled to make up for the disappearance of two small and obsolete refineries. This was accomplished in 1977, with the new first-phase unit working on Mangyshlak oil delivered by tanker. This has created problems, because it is necessary to receive and refine the local high-quality oil and Mangyshlak sulphurous oil separately. A large new catformer of capacity 1 mn tons a year was installed by Litwin at a cost of $50 mn in 1981, and the plant's catcrackers have been modernised and re-equipped. A new first-phase plant was completed in 1979, bringing the refinery's capacity to 18 mn tons a year.

Batumi (Georgia)

This small refinery dating from 1930 handles oil brought by rail from Baku and by pipeline from the new Samgori oilfield. It has a capacity of 6 m.t. a year.

Tbilisi (Georgia)

Tbilisi is a small, old and obsolete refinery of capacity 3 m.t. a year.

Novo-Gorki (Volga Vyatka Region)

The Novo-Gorki refinery, situated at Kstovo, came on-stream in 1958. Its capacity was raised to 24 m.t. in time to receive oil from Tyumen through the Surgut-Polotsk pipeline, which reached Gorki at the beginning of 1980.

Kremenchug (Ukraine)

After coming on-stream in 1966, Kremenchug was supplied with oil by rail until 1974, when the Michurinsk-Kremenchug pipeline was completed. In 1978, another pipeline, from Lisichansk, was laid; this is part of the system distributing Tyumen oil throughout the Ukraine, and it enabled the capacity of the refinery to be increased to 15 m.t. The refinery has a large secondary refining capacity, and a large share of its products are distributed by pipeline to Cherkassy, Kiev and Chigirin power station. A lube additive plant came on-stream in 1981.

Kherson (Ukraine)

Dating from 1960, Kherson refinery is supplied by a pipeline from Kremenchug bringing Tyumen oil and another from Poltava bringing local Ukrainian oil. It has a capacity of 9 m.t. a year.

Lisichansk (Ukraine)

Lisichansk is one of the USSR's newest refineries. It began operating in September 1976 on crude received through a pipeline from Krasnodar. In 1977 a major pipeline of diameter 1,220 mm and 1,089 kms long from Kuibyshev started bringing Tyumen oil. The second primary unit was completed in October 1979, and in 1980, the first part of the second stage was commissioned, bringing the refinery's capacity to 18 mn tons a year. In 1980, a new catformer of 1 mn tons a year was introduced, and 1981 saw the start up of a 2 mn tons a year hydrofiner with a high degree of automation, said to be operated by only 6 workers. It is estimated that diesel from the hydrofiner will enable the working life of lorry engines to be prolonged by 15 to 20%. A distinctive feature of the refinery is its ethylene production unit, which began operating in 1978. Eventually it will produce 300,000 tons a year and will be connected by a 36 kms pipeline to the chemical plant at Severodonetsk, where the ethylene is turned into polyethylene.

Odessa (Ukraine)

Odessa has a small refinery, now working on Tyumen oil delivered by the final stage of a 4,500 kms system originating in Western Siberia. It has a capacity of 6 m.t. a year.

Drogobych (Ukraine)

This is a small primary refining plant kept alive by oil railed from the new Kaliningrad field. It has a capacity of 3 m.t. a year.

Lvov and Nadvornaya (Ukraine)

These two small refineries with a joint capacity of 6 m.t. a year work on local oil from the declining Western Ukrainian field.

Perm (Urals Region)

The Perm refinery began producing in 1957 and has since been expanded on several occasions, making it one of the USSR's largest refineries with a capacity of 24 m.t. It is notable for the production of petro-chemicals with a new plant for producing ethylbenzene and styrene from low-grade oil starting up in 1975, and an ammonia and carbamide complex with an annual capacity of 1 m.t. of fertilisers coming on-stream in 1980.

Krasnokamsk (Perm oblast, Urals Region)

This small refinery, located near Perm, uses local oil and has a capacity of 6 m.t. a year.

Orsk (Urals Region)

Orsk refinery, built before the last war, processes Bashkir and Tyumen oil which arrives through a pipeline from Salavat. It has a capacity of 12 m.t. a year.

Ufa (Bashkir republic)

Ufa refinery was built in 1936 and has been reconstructed and expanded several times. It uses local oil and is part of a complex including the Ufa 22nd Congress and Novo-Ufa Lenin refineries and a number of petro-chemical plants. The total capacity of the complex is 46 mn tons a year, and its product output rose by 22% over 1976-1980.

Ufa 22nd Congress

Unlike the other two refineries in the complex, which use local Bashkir oil, Ufa 22nd Congress has been working on Tyumen oil since 1970. With the Urals-Volga oilfields beginning to decline, and with the Urengoi-Chelyabinsk gas pipeline serving many needs formerly covered by fuel oil, the emphasis in recent years has been on installing secondary capacity, like a 1 mn ton a year catformer built in 1981 by Litwin and Technip of France. However, it is said that the refinery has not been working well for a number of years, and urgent measures must be taken during the Eleventh FYP period.

Novo-Ufa Lenin

A new catalytic cracker was installed in 1973, and a catformer intended to produce a larger proportion of the USSR's high-octane petrol for Ladas was built in 1976. The bitumen plant is the chief supplier for a large part of the Volga region.

Saratov (Volga Region)

This refinery has a large Czech-built catformer which was installed in 1973 and produces high-octane petrol for Volgas. The capacity of the refinery is about 15 million tons.

Volgograd (Volga Region)

Volgograd refinery processes local oil from the Lower Volga region and Mangyshlak oil delivered by tanker. It is technologically advanced with a large volume of secondary refining capacity, and produces over 50 different products. A carbamide deparaffinisation plant was introduced in 1980, and this portends increased supplies of oil from the Mangyshlak and Buzachi peninsulars. Its current capacity is 12 mn tons a year. A parex plant produces 120,000 tons a year of liquid paraffin for the Svetloyarsk Plant of Vitamin Concentrates.

Novo-Kuibyshev (Volga Region)

After coming on-stream in 1949, this refinery has become one of the world's largest with a capacity of 30 m.t. It refines oil from local fields and from Tyumen and Mangyshlak. It has traditionally been a research centre for the oil refining industry, and a new type of fully automated parex plant which can be operated by only 12 people is being tested there. It will produce paraffin for the synthetic detergent and animal feed industries. A 1 mn tons a year catformer was installed in 1981.

Kuibyshev (Volga Region)

Kuibyshev is an old primary processing refinery with a capacity of 6 m.t. It sends some of its residual fuel oil to Novo-Kuibyshev for further refining.

Syzran (Volga Region)

This is a small refinery of capacity 6 m.t. A new parex plant which produces 24 tons of paraffin per 100 tons of raw material compared with the usual 14 tons has recently come on-stream.

Ishimbai-Salavat (Volga Region)

The Ishimbai refinery, built in 1934, has been incorporated into a complex with the Salavat petro-chemical plant. It processes oil from local Bashkir fields, gas condensate and casinghead gas. It has a platformer producing 93 octane petrol, a Czech-made hydrofiner for processing diesel and equipment for producing dearomatized kerosene which is used for the manufacture of detergents at Sumgait near Baku. The refinery has a capacity of 12 m.t.

Nizhnekamsk petro-chemical combine (Volga region)

This is one of the newest refineries in the USSR, and the first in the Tatar republic. It is served by a new pipeline from Almetyevsk through Noberezhnye Chelny, which brings Tyumen oil arriving through the Samotlor-Kuibyshev pipeline. The primary refining facility came on-stream in April 1979, and three or four different petro-chemical processing facilities are installed each year. The refinery now has a capacity of 12 m.t.

Omsk

The Omsk combine dates from the early 1950s, and it has been extended many times since then. At first it worked on Volga oil sent through the Tuimazy-Omsk pipeline, but it is now fully converted to the refining of Tyumen oil. In 1974, the first plant in the USSR for making vehicle transmission oil was installed, and secondary capacity was further increased with the introduction of a hydro-cracker in 1978. The refinery was extended further in 1979 and 1980 when a 1 mn tons a year catformer was installed by Litwin at a cost of $36 mn, and a petroleum coke plant is now being built. Omsk refinery now has a capacity of 30 mn tons a year.

Angarsk (Eastern Siberia)

This refinery is part of the Angarsk Petro-Chemical Combine and has a capacity of 30 m.t. It uses Tyumen oil delivered through the pipeline from Samotlor to Anzhero-Sudzhensk where it joins the Trans-Siberian Pipeline. It supplies most of the Eastern Siberian region with oil products and the rapidly growing local petrochemical industry with feedstocks. During 1981 its desulphuriser and catcracker were modernised, and the pipeline for carrying ethylene from its 300,000 tons a year EP-300 ethylene facility to Zima completed.

Komsomolsk (Far East)

The small Komsomolsk refinery was built in 1935 with a capacity of 1 million tons a year. It refines Sakhalin oil arriving through the Okha-Komsomolsk pipeline, and in 1979 its capacity was trebled in time for the completion of a second string of the pipeline. The refinery is not yet working at capacity; this will happen when the Mongi-Pogibi pipeline on Sakhalin is completed. Its capacity is currently 3 m.t.

Khabarovsk (Far East)

Built in 1950, the Khabarovsk refinery of capacity 9 m.t. refines Tyumen oil sent by pipeline to Irkutsk and then railed to Khabarovsk. It now produces 16 types of product, completely satisfies the Primorskii Krai's requirements for petrol and diesel, and produces fuel oil for export.

Gurev (Kazakhstan)

This small refinery was built in 1947 and enlarged in 1971. It processes local oil from the Emba field and heavy oil from Mangyshlak. Its capacity is now officially said to be 3 m.t. A coking plant producing coke for the non-ferrous metallurgy industry started up in 1980.

Pavlodar (Kazakhstan)

Pavlodar refinery was completed in 1978, three years behind schedule. It is fed by the Omsk-Pavlodar pipeline which delivers Tyumen oil. A secondary primary refining stage and a bitumen producing complex of capacity 0.5 m.t. a year started up in 1980, and in 1981 a gas bottling plant began to collect and process gas which was previously flared. The refinery now covers bottled gas demand from most of Kazakhstan. A 1 mn tons a year catcracker is currently under construction, and the refinery now has a capacity of 12 mn tons a year.

Krasnovodsk (Central Asia)

Krasnovodsk refinery is famous for its new coking plant, the first of its type in the USSR. It was manufactured by Creusot-Loire of France and delivered in 1974. The plant started producing high-quality electrode coke for the steel and aluminum industries in 1977. Another such plant has since been installed and plans have been drawn up for a thermo-cracking installation designed to produce needle-like coke. The refinery works on local oil, and has been operating at much less than its rated capacity of 12 mn tons a year in recent years because of the decline in Turkmen oil production.

Fergana (Central Asia)

Built in 1959, this small refinery was extended in 1972. It is supplied by local oil from the Uzbek and Tadzhik fields. Its capacity is about 4 mn tons a year, but it is considerably underutilised and has a poor record for efficiency.

According to the Eleventh FYP, the period 1982-1985 should see the completion of the first stages of the Achinsk, Chardzhou and Chimkent refineries, and new capacity should be added at Lisichansk and Pavlodar. All these new units should be of 6 mn tons a year each. The Mazheikiai refinery will be extended by 18 mn tons a year, and Khabarovsk by 3 mn tons. It can be estimated that primary refining capacity should rise to 612 mn tons a year by the end of 1985. The Groznyi and Baku complexes will be reconstructed and expanded to enable them to process more oil from Siberia and the Buzachi field of Kazakhstan.

The most efficient plants in 1981 were Moscow, Isimbai, Pavlodar, Gurev, Novopolotsk, Kherson, Yaroslavl and Tuapse. They have an aggregate first phase capacity of 98 mn tons a year. Four refineries failed badly to meet their targets. These were Fergana, Osk, Groznyi Sheripov and Krasnovodsk, with a total installed capacity of 34 mn tons.

REFINING TECHNOLOGY

Improvements in refining technology have proceeded in two directions: a larger size of refinery and refining units, and a greater depth of processing with a concomitant improvement in the quality of products.

Immediately after the last war, the capacity of refining plants was very small. First-phase plants of 600,000 tons a year, thermal crackers of 450,000 tons and catcrackers of 250,000 tons a year were standard refining equipment. By 1982, domestically produced first-phase units of 6 mn tons had been standard for 17 years, and some of 8 mn tons had been installed. Catcrackers and catformers of 1 mn tons a year and hydrofiners of 2 mn tons were operating at several of the larger refineries. During the period 1976-1980, the volume of catformer capacity accounted for by units of 0.9 mn tons a year and more rose from 27 to 42 per cent, while the volume share of hydrofiners of 1.5 mn tons a year and more grew from 20% to 30%. The standard size of new vacuum distillation units is 5 mn tons a year, and modern Soviet lube plants have capacities of 0.38 mn tons a year.

As long ago as 1974, first phase plants of 12 mn tons, catcrackers of 1.5 mn tons and catformers of 1.2 mn tons had been designed, but the Russians appear to be sufficiently happy with their standard equipment, and particularly large or advanced units have been imported. These include catformers of 1 mn tons capacity which have been built at several refineries by Litwin.

During the six years 1975-1980, 871.8 mn roubles of oil refining equipment was imported. This included 394 mn roubles from East Germany (mostly small crackers and parex liquid paraffin units), 207.1 mn roubles from Japan, 80.9 mn roubles from Czechoslovakia, 79.9 mn roubles from France and 40.8 mn roubles from Romania.

The most pressing problem facing the refining industry at the moment is the need to upgrade its products. As noted in the introduction, fuel oil accounts for more than 50 per cent of oil products by volume, and it is used for purposes which could be met more rationally by natural gas. The arrival of the two natural gas pipelines in the Urals region, for example, has enabled a large number of industrial and domestic boilers which previously burned high-sulphur fuel oil from the Ufa complex to switch to natural gas. This has improved the environment in Urals cities and lowered costs, but has also left the refineries with a surplus of fuel oil on their hands. This problem can only be solved by the installation of secondary refining equipment.

It has been argued that the Ufa refineries are not doing enough to overcome this problem. The programme of construction and reconstruction for the 1981-1985 period anticipates a decline in the share of fuel oil (mazut) in the product mix from 41.5 per cent to 37.0 per cent, but this is regarded by the Bashkirneftek-himzavod Association as insufficient. By making a greater use of vacuum distillation facilities, it is said that "hundreds of millions of roubles" worth of light and medium products could be obtained without any increase in the throughput of crude. In the long term, it is believed that the Ufa complex can increase the depth of refining to 85%, and a plan is being drawn up, involving the installation of visbreakers, catcrackers, hydrofiners, coking plants and lube plants.

In other regions, refineries face the problem of how to meet a growing demand for light products from a declining supply of crude. This is the case at Baku, where a considerable amount of money has been spent on rationalising the city's refineries by installing more secondary equipment such as the catformer and catcracker at Novo-Baku Vladimir Ilyich.

However, the Baku refineries are not meeting their plan targets because of a shortage of high quality crude. This problem can only be rectified when the Groznyi-Baku oil pipeline is completed and Siberian crude begins to arrive.

Current Soviet policy is to raise the share of light and medium products from 50 to 60-65 per cent, but at the present time existing secondary units at some refineries are underutilised. Particularly high rates of output of light products, and a corresponding fall in the share of fuel oil, were envisaged for 1980, and this was to be achieved largely through the commissioning of catformers at Kuibyshev and Kirishi. The output of liquid paraffin intended for the production of animal feed additives and synthetic detergents was planned to increase considerably. The parex method of paraffin production was installed at five refineries: Ukhta, Kirishi, Syzran, Novo-Kuibyshev and Mozyr.

As the depth of refining has grown, the quality of oil products has improved. It was planned to raise the share of high-octane petrol in total petrol output from 50 per cent to 74 per cent during the Ninth FYP period (1971-1975), but it reached 84.8 per cent by 1975. By the end of 1979, 93.8 per cent of petrol production was high-octane. At this point, the definition of "high-octane" was revised, and the Eleventh FYP has called for the share of "high-octane benzine" to be raised from 64% in 1980 to 87% in 1985 and 96% in 1990. The improvement of diesel fuel has been a particularly important achievement. Soviet economists have estimated that diesel with a 1 per cent sulphur content wears an engine out after 57,400 kms, while diesel with a 0.2 per cent content will keep it running for 88,700 kms. The share of low-sulphur diesel fuel (i.e. sulphur content of less than 0.5 per cent) has achieved 95.8 per cent, including 47.2 per cent with a sulphur content of less than 0.2 per cent. The Eleventh FYP expects all diesel to be low sulphur in 1985. The installation of barbamide deparaffinisation plants at a number of refineries has enabled diesel with very low freezing points for the northern regions to be produced. A further aim of the Eleventh FYP is a big increase in the production of lube additives, the use of which is expected to effect savings of 350 mn roubles for the economy.

The introduction of modern, more efficient equipment has lead to a significant reduction in the losses of oil and oil products during the refining process. In 1980, a total 5.7 mn tons were lost during processing, or 1.25 per cent of throughput, and this compares with 1.51 per cent in 1975. It is planned to reduce losses to 1.1 per cent in 1985.

Some refineries, such as Kirishi, Yaroslavl, Mozyr, Novopolotsk and Ryazan have already reduced the loss rate to less than 0.7 per cent a year. These are all comparatively modern plants, and the replacement of obsolete equipment

at other refineries should enable the rather high national loss rate to be
reduced. In particular need of attention are the Fergana and Krasnovodsk plants
of Central Asia, where losses amount to 3-4 per cent of throughput.

Oil refineries are important users of their own products and the Eleventh FYP
hopes to reduce such "losses" by improving the use of secondary energy resources.
In 1980, they produced 11 mn G-cals of heat and steam, and the planned installation of 150 boiler utilisers should permit this to rise to 22 mn G-cals in
1985. Together with the planned reconstruction of heat exchangers at 12 first
phase plants and the modernisation of a number of furnaces, it should be possible
for the refinery sector to save almost one million tons of fuel over 1981-1985.
The planners believe that a further 0.5 mn tons can be saved by the use of
refinery gases which are still flared at some refineries.

GAS REFINERIES

The production of casinghead gas was planned to grow from 28.6 bn cu m in 1975
to 43-45 bn in 1980, but in fact rose to only 37.5 bn cu m. The introduction of
12 bn cu m of gas refinery capacity during 1976-1980 has enabled the rate of
utilisation of casinghead gas to be raised from 57.8 per cent in 1975 to 70
percent in 1980 when the remaining 30 per cent was either flared or pumped back
for the gaslift extraction of oil.

The Tenth FYP called for a utilisation rate of 80-85 per cent to be achieved in
1980, and this required six new gas refineries to be built in Western Siberia-
Nizhnevartovsk Nos. 2, 3 and 4, Yuzhno-Balyk, Belo-ozersk and Surgut. It
would also be necessary to raise the rate of gas utilisation in the Orenburg,
Emba and Central Asian oilfields. With the failure to fulfill the targets of
the Tenth FYP, the 85% utilisation target has been put back to 1985 according to
the Eleventh FYP, and the increase from 70% in 1980 should allow the output of
LPG to rise by 5 mn tons.

Apart from a 0.5 bn cu m extension to the Kazakh refinery in 1977, and the
completion of the 2 bn cu m facility at Usa (Komi republic) in 1980, all new gas
refinery capacity has been installed in Western Siberia.

Table 17

Capacity of Gas Refineries in Western Siberia, bn cu m, beginning of year

	1975	1976	1981	1982
Nizhnevartovsk	2	8	8	8
Pravdinsk	0.5[a]	0.5	0.5	0.5
Yuzhno-Balyk	-	0.5	0.5	0.5
Belo-ozersk	-	-	4[c]	4
Surgut	-	-	-	4[d]
Total	2.5[b]	9	13	17

a Muravlenko & Kremneva "Sibirskaya Neft" p. 90   b Neftyanoe khozyaistvo
1979 No. 4   c Izvestia 10/8/80   d Pravda 13/12/81

Construction of the Nizhnevartovsk gas refinery began in 1970, and by 1975 the first of the planned four units of 2 bn cu m capacity a year each was operating, in addition to the small plant at Pravdinsk. By the end of 1978 the last unit had been installed, although it did not begin operating until mid 1979. Total output of casinghead gas from Western Siberia exceeds refinery capacity because it includes gas from the Fedorovsk gas cap.

In 1980, Nizhnevartovsk reached its projected capacity, enabling 70 per cent of all gas from Samotlor to be utilised, and the Belo-ozersk plant of capacity 4 bn cu m/year was completed in 1980. Samotlor gas is now fully utilised, and the Belo-ozersk refinery will be brought to full capacity by processing gas from the Varyegan group of deposits. Belo-ozersk is connected with Nizhnevartovsk by a 70 km concrete highway. It was built by the Megiongazstroi trust and was completed in 2½ years, a year behind schedule due to shortages of materials and equipment.

In 1980, the first phase of 2 bn cu m a year was completed at the Surgut refinery, which had been built to utilise the casinghead gas from Fedorovsk and other local oil deposits, and the second stage started up at the end of 1981.

During the Eleventh FYP period, new refineries will be built at Lokosovo (to serve the new Middle Ob deposits between Nizhnevartovsk and Surgut), Tarko-Sale (for the large new deposits due to start up between Kholmogorsk and Urengoi) and Krasnoleninsk, for deposits on the lower reaches of the River Ob. By 1985, it is anticipated that all Western Siberian casinghead gas will be fully utilised.

Technical improvements to Soviet gas refineries have centred on improving the low temperature absorption and condensation processes. Absorption has high capital costs and requires comparatively large amounts of energy, and the VNIPIgazpererabotka Institute has been elaborating different ways of raising its efficiency. The low-temperature condensation process is simpler, requiring temperatures of -35 to -45 degrees and pressures of 35 to 40 kgs per sq cm. The Russians have raised the pressure to 59 kgs/sq cm and lowered the temperature to -82 degrees for their Nizhnevartovsk, Belo-ozersk and Surgut plants, so as to achieve a propane recovery rate of more than 90%, but VNIPIgazpererabotka has subsequently devised ways of obtaining the same result with pressures of 35 to 37 kgs/sq cm.

The USSR's second biggest gas refinery, at Minibaevo in the Tatar republic, uses a low-temperature condensation and rectification process using a double-cascade cooling cycle employing propane and ethane. With a temperature of -60 degrees, the facility is more efficient than other processes. The importance of absorption plants is declining in the USSR; they are used mainly for the processing of lean gases containing 3 to 5 per cent heavy fractions.

The principal aim of current research is to double the standard size of refining units from the current 1 bn cu m a year, largely by raising the size of compressors and cooling equipment.

CHAPTER 5

The Resource Base

RESERVES OF OIL AND GAS

Exploration for oil and gas is generally carried out by organisations belonging to the Ministry of Geology, while organisations under the Ministries of Oil and Gas confine themselves to appraisal drilling, and carry out exploration drilling only in heavily worked regions where they are well established.

Although disagreements and rivalry can arise in some regions, relationships between the Ministry of Geology and the fuel-producing ministries are generally close. This is because the long term plan of the Ministry of Oil (for example) involves the need to prepare new oilfields for future development well in advance. It is often assumed in the West that the ministry's plans extend only as far as the end of the current FYP period. In fact there is evidence that the pre-planning of development of new fields is carried out for up to 15-20 years in advance, with the Western Siberian Research Institute of the Geology of Oil and Gas occupying itself with the formulation of annual plans for the preparation of reserves of oil and gas up to the year 2000.

Ilchenko and Yasin of the Research Institute of Organisation and Economics of the Oil and Gas Industry have provided an insight into official policy.[1] They give the example of a hypothetical new oilfield where the first stage of development, that of geophysical work, has suggested the presence of a potential 10 bn tons of oil. Anticipated well-yields and capital investment costs have suggested that 450 mn tons of oil will be proved up to the "proved + probable" standard by the time the field is ready for exploitation. According to their calculations, three years would be required for the second stage of development (wildcatting, involving the discovery of at least one significant deposit), 5.3 years for the third stage (drilling appraisal wells) and one year for the final stage of creating the necessary infrastructure for exploitation. Thus some 9.3 years would be required before the field would come on-stream.

The length of this period may vary considerably, depending on the size of the field, its well-yields and its proximity to established pipelines. But it is almost certain that the USSR is in the process of preparing fields at the present time which are not expected to come on-stream until beyond 1990.

While the sizes of Soviet reserves of gas and coal are known to be larger than those of any other country, and are adequate for decades of exploitation, it is the volume of oil reserves which excites most interest in the West. Soviet oil reserves have been a state secret since 1947, and Western estimates have varied between the sublime and the ridiculous, or between 22 and 6 billion tons of proved plus probable reserves.

It can be estimated from data provided by M.V. Feygin[2] that proved + probable reserves amounted to 12.4 bn tons in 1972, of which 4.9 bn were proved. Some 849 small deposits contained 1.67 bn tons, 132 medium deposits had 2.53 bn tons, 24 large deposits had 1.52 bn tons, and 18 giant and supergiant deposits with over 100 mn tons each contained a total of 6.67 bn tons.

---

1 Ekonomika Neftyanoi Promyshlennosti 1979 No. 1 2 Neftyanye resursy, metodika ikh issledovaniya i otsenki, M V Feygin Moscow 1974 p. 10.

Feygin also gives the proportions of total reserves in both proved and proved + probable categories found in Paleozoic, Mesozoic and Cainozoic structures. It can be estimated from his data that 82 per cent of the increase in reserves over the period 1961-1972 has come from Mesozoic structures, which are found mainly in Western Siberia. Such oil accounts for almost half the total volume of reserves in 1972, and, as Robert Campbell says, "Half of the explored reserves of the Soviet Union are now in this region".[1] If all the new reserves found in Mesozoic structures are attributed to Western Siberia, then it would have had a total of 5.7 bn tons of proved + probable oil in 1972.

Since 1972, reserves have continued to grow, although more slowly than before. Academician Trofimuk says that, in spite of the rapid increase in oil production in Western Siberia, the reserves-to-production ratio is increasing, but very slowly.[2] Krylov says that the share of proved reserves in total proved + probable reserves is falling,[3] in spite of the large amount of appraisal drilling taking place at currently operating fields, and this suggests that comparatively large reserves of probable oil are being found.

During the Tenth FYP period, 370 hydrocarbon deposits were discovered as well as 330 pools at existing deposits. This included 220 new oil deposits and 280 oil pools. During the same period, 12,70 areas (ploshchady) were explored by seismic field parties and prepared for exploratory drilling.

There is not much dispute about the size of proved + probable reserves of gas. The monthly publication of the Ministry of Gas, Gazovaya Promyshlennost, used to publish reserve figures annually until they reached such an immense size that they excited little further interest.

Table 18

Soviet Proved + Probable Reserves of Gas, bn cu m, beginning of year.

	1960	1970	1975	1979	1980	1981
USSR	2,202	12,092	24,600	32,100	33,300	35,000
Western Siberia	78	7,430	16,100	23,900	25,000	27,000
Central Asia	569	2,034	3,300	3,200	3,500	3,600

Sources: Gazovaya Promyshlennost (various issues) and authors estimated.

Annual plans for the discovery of new gas reserves have been consistently over-fulfilled since the late 1960s. This has been largely due to the phenomenal discoveries in Western Siberia, and more recently to the discovery of large new deposits of gas in the Uzbek reefs of Central Asia.

EXPLORATION OF WESTERN SIBERIA

The search for oil and gas in the region is assisted by three geological institutes in the city of Tyumen, employing a staff of more than 2,000.

For the purposes of carrying out exploration in a rational, ordered manner, Western Siberia has been divided into 31 zones. In twelve of these zones,

1 Soviet Economic Prospects in the Seventies, Joint Economic Committe of the US Congress, Washington 1973, p. 46. 2 Platts Oilgram News 3/7/80 3 EKO 1980 No 1

regional exploration projects are now under way; a further eleven projects are being studied, but exploration will not take place in these zones until after 1990.

Almost all exploration work in Western Siberia is carried out by Glavtyumengeologiya of the Ministry of Geology, with Glavtyumenneftegaz of the Ministry of Oil and Tyumengazprom of the Ministry of Gas responsible only for a small volume of appraisal drilling. In fact the Ministry of Oil's share of exploration drilling has declined from 10 per cent in 1975 to about 2.5 per cent by 1979.

A common Western criticism of Soviet oil exploration work has been that the explorationists have been attempting to meet their plan targets by the concentrating on appraisal drilling at existing deposits where conditions are easiest. In fact, exploration teams have been moving further and further away from the established centres of the Siberian oil industry along the Ob River, and this is reflected in deteriorating indices relating to commercial drilling speeds, well completion times, etc. Between 1975 and 1978, the commercial drilling speed fell from 1,230 to 1,005 metres per rig per month, labour productivity of wildcatters has fallen from 44.5 to 37.3 metres per driller, and the average well completion time has risen from 156 to 216 days.

Glavtyumengeologiya consists of five production associations which, between them, administer 28 exploration expeditions. The most important of these, in terms of new oil and gas discoveries, belong to the Yamalneftegazgeologiya Association, based on the Yamal-Nenetskii Autonomous Okrug. They include the Karsk expedition, which is currently engaged in proving up the supergiant Bovanenkovsk gas deposit on the Yamal Peninsula, and the Tazovsk expedition which discovered Urengoi in 1966 and is now concentrating on exploring the Russkoe heavy oil deposit and drawing up detailed maps of Urengoi and Yubileinoe gas deposits. The Urengoi expedition was specially established for appraisal drilling at Urengoi, but has also been discovering other gas and oil deposits in the Urengoi area, and the Nadym expedition has been drilling to deeper than 5,000 metres in the area of the Medvezhe deposit, looking for deeper payzones. The most recent expedition is the Tambeisk, created in 1980 to work the northern shore of the Yamal Peninsular.

The explorationists are assisted by a special geophysical organisation, the Yamalneftegazgeofizika trest. During the 12 years of its existence, it has surveyed 600,000 sq kms and prepared 11 structures for exploratory drilling, which lead to the discovery of 47 oil and gas fields including giants like Urengoi, Yamburg, Bovanenkovsk, Venga-Yakhinsk and Novoportovsk. During 1976-1980, it explored 15,558 sq kms including 3,830 sq kms in 1980 (the plan targets were 14,560 and 3,600 sq kms), and prepared the Yety-Purovsk, Vostochno-Tarkosalinsk, Zapadno-Seyakhinsk and Tota-Yakhinsk areas for exploratory drilling. It has 20 field parties, including 5 new ones set up during the 1981-1982 season at Nadym, the Gyda Peninsular, Se-Yakha, the Taz River Basin and the Purovskii Raion.

The three most important associations operating in the oil-producing region are Megionneftegazgeologiya, which is based in Megion and is responsible for exploration of the Nizhnevartovsk field, the Obneftegazgeologiya association of Surgut, which discovered Fedorovsk, Kholmogorsk and Muravlenkovsk and is now working on 4 new areas, and the Pravdinsk Association which is concentrating on the proving up of the Krasnoleninsk field west of Surgut.

During 1976-1980, Megionneftegazgeologiya overfulfilled its target for new oil reserves by 4 per cent and discovered 17 deposits. In 1980, it drilled 89 wells to a total depth of 350,000 metres (the plan was for 87 wells of 324,000 metres) accounting for 33.8% of all exploratory drilling in Tyumen oblast. The Agansk expedition of the Megion association has been particularly successful during the last five years in its work on the Varyegan group of deposits to the north of Samotlor, and in 1980 the appraisal of the Varyegan and Severo-Varyegan deposits was completed.

For several years now, a dispute has been simmering between explorationists and production workers over Western Siberia's oil-producing potential. The geologists maintain that Tyumen oblast has sufficient resources to keep production rising steadily for many years, and therefore production rates can be considerably stepped up. This view was expressed as recently as September 1979 and it is argued that the onus falls on development drillers to increase the number of development wells drilled per year.

The speed at which Tyumen's oil industry has grown has surprised planners. It was thought in 1970 that 1980's output might be as little as 230 mn tons, and a decision was taken to reduce exploration rates because it was believed that sufficient oil had already been discovered to sustain the anticipated development rates. The principal reason why development has been faster than planners believed possible is that Glavtyumenneftegaz has successfully kept output targets as low as possible so as to enhance its chances of plan fulfillment, and thereby maximise its bonuses.

It is no coincidence that, in spite of all the uncertainty surrounding planning for such a new and rapidly growing region as Tyumen oblast, annual plans since 1970 have been met exactly, year after year. Had they wanted, Tyumen oil workers could have produced far more oil. But in the long term, it pays them to meet targets year after year and earn a regular fulfillment bonus rather than substantially overfulfill one year, and then face difficulties in meeting upwardly revised targets during the rest of the FYP. In other words, five fulfillment bonuses are worth more than one overfulfillment bonus.

The cutback in exploration drilling from a peak of 500,000 metres in 1967 to 384,000 in 1972 was accompanied by a freeze on the creation of an infrastructure necessary for the drillers. Settlements, roads, communications sytems and supply depots were not created, and it consequently became more and more difficult for drillers to fulfill their tasks. In 1974, only 79.9 per cent of the planned volume of exploratory drilling was carried out, and in each of the five years 1971-1975, drillers failed to fulfill their targets. But as the finding rate was substantially higher than expected, this enabled the plans for discovery of new reserves to be met each year. It is suspected, however, that the higher-than-anticipated finding rate was due to the concentration of exploration work on the most promising areas where substantial volumes of oil had already been discovered.

Table 19

The Finding Rate in Western Siberia
(% of plan target)

	Exploratory Drilling	Discovery of oil	Finding rate
1971	97.4	117.0	120.1
1972	96.4	110.0	114.1
1973	78.9	103.7	118.0
1974	79.9	103.6	129.7
1975	97.5	103.1	105.7

Source: Columns 1, 2: Ekonomika Neftyanoi Promyshlennosti 1977 No. 2. Column 3: Calculated from columns 1 and 2.

Although plan targets for new reserves were met, the actual amount of new oil discovered fell in 1971 over 1970, and not until 1975 did it begin to exceed 1970 levels significantly.

Table 20

Discovery of Oil
in Western Siberia by Glavtyumengeologiya

	% of plan	% of 1970
1970	-	100
1971	117.0	66.7
1972	110.0	93.2
1973	103.7	102.0
1974	103.6	112.0
1975	103.1	125.0

Source: Ekonomika Neftyanoi Promyshelennosti
           1977 No. 2

In 1975, only 728,000 metres of exploratory wells were drilled; this included
61,000 metres undertaken by Glavtyumenneftegaz and 667,000 by the Ministry of
Geology of which 556,000 were drilled in Tyumen oblast by Glavtyumengeologiya-
93,000 in Tomsk and 18,000 in Novosibirsk oblast. Exploratory drilling amounted
to little more than a quarter of development drilling.

But as the reserve to production ratio in Tyumen oblast grew very slowly, it
became apparent that the rate of exploration must increase, so targets were
raised. But little was done to improve the infrastructural shortcomings, and
the targets were not achieved. In 1976 only eight of the 24 exploration teams
belonging to Glavtyumengeologiya met their targets, and the plan for increasing
oil and gas reserves was underfulfilled. In 1977, the association finished the
year 48 days behind schedule and drilled only 533,000 metres. Rig erectors
were blamed for this, with only 13 rigs being erected in November, for example,
compared with the planned 28.

Things improved in 1978 when drilling increased by nearly 56 per cent to 831,000
metres. The plan for new oil discoveries was fulfilled and 30 new structures
were examined.

Glavtyumengeologiya's operations were planned to expand dramatically after 1978.
The number of drilling crews was expected to grow from 56 in 1978 to 100 by the
end of 1980, and each crew was expected to drill 13,500 metres in 1980. The
emphasis remained on the easier structures in the Middle Ob, the Gyda Peninsula
and the Nadym-Pur and Pur-Tazov regions, and on finding more pools at existing
deposits. The non-structural prospects such as stratigraphic traps will remain
neglected until the quality of seismic equipment improves and the seismologists
become more experienced.

The annual plan for 1979 was to drill 1 mn metres of wells, erect 400 rigs and
test about 300 exploration wells. The plan was drastically underfulfilled,
mainly because the severe winter prevented rig-erectors from getting out to the
prospective areas before summer set in and the terrain became an impassable
swamp. Only 763,000 metres were drilled. The 1980 plan called for an increase
of 50 per cent over 1979 to give a target of 1,150,000 metres, and while the
target was underfulfilled, with only 1,037,000 metres drilled, an increase of
35.9 per cent over 1979 was achieved. Nine deposits and 23 pools of oil and
gas were discovered. Some 30 new areas were explored.

In 1981, 1,150,00 metres were drilled (10.9 per cent more than 1980), 36 new
areas were explored, and 10 new deposits were found, including 9 of oil and

one of gas condensate. During the Eleventh FYP period, Glavtyumengeologiya plans
to drill 10 mn metres of exploration wells, or more than the total so far drilled
in Western Siberia. This will consist of 3 mn metres in the Yamal-Nenetskii
gasfield area and 7 mn metres in the Khanti-Mansiiskii oilfield area. The
Yamalneftegazgeologiya Association plans to drill 400,000 metres in the Yamal-
Nenetskii region in 1982, and 1 mn metres in 1985. Given the large volume of
gas reserves already discovered, it may seem more logical to direct some of
the Yamal drilling crews to the more southerly oilfields. However, it is believed
that large oilfields will be discovered in the Yamal-Nenetskii region in the
1980s .

By 1980, over 200 deposits of oil and/or gas had been discovered in Tyumen
oblast, including over 500 pools of oil, gas or condensate. Of these, 40
oil and six gas deposits were producing by the end of 1980, and a further 27
oil deposits are to come on-stream during 1981-1985.

During 1971-1975, 22 new oil deposits were discovered in the Khanti-Mansiiskii
oilfields, several of which are now producing. It is said that exploration
work on the middle reaches of the River Ob, where the biggest oilfields are
located, has reached the concluding stage, and drillers are moving further north.
During the 1970s, wells drilled into Jurassic and chalk strata in the far north
of Tyumen oblast gave flows of oil at the following gas deposits: Arkticheskii,
Vyngapur, Gubkin, Zapolyarnyi, Medvezhe, Samburg, Urengoi, Yubileinyi, Yarinersk
and Yamburg. During the current plan period, the search for oil will concentrate
on Urengoi and the Yamal Peninsula.

After the cutback in exploration work in the early 1970s , the discovery rate
is picking up again with 50 new fuel deposits being found during the five years
1976-1980. During this period, a large amount of work was carried out on the
Yamal and Gyda Peninsulas in the far north, in the Tazovsk area (i.e. in the
vicinity of Russkoe deposit), in the Tarko-Sale area, in the region of the
Kogolymsk and Kholmogorsk deposits between Surgut and Tarko-Sale, along the
Agan river in the vicinity of the Varyegansk group of deposits, and to the west
of Surgut in the Krasnoleninskii area.

A number of large gas and gas condensate deposits were found on the Yamal Peninsula. They include the Malyginskaya field on the northern coast of the Yamal
Peninsula, and Severo-Tambeisk, located in its north-east corner, 1,190 km
north of Surgut and 200 km north-east of the Yamal supply base at Mys Kharasavei.

Drilling began on the Gyda Peninsula in April 1977 when the first well was drilled
at the Gydanskaya Ploshchad prospect. Nothing much was found until mid-1979,
when the promising deposit of Utrenneye was discovered near the village of
Tadibeiyakha on the west coast of the peninsula. It gave both oil and gas
condensate.

Oil is also being found in Paleozoic strata at very great depths. The first deposit,
Malo-Ichsk, was discovered in 1974, and this was followed shortly afterwards by
discoveries in the Urmansk and Nizhne-Tabagansk regions of Tomsk oblast where
commercial flows were obtained. More recently, oil has flowed from depths of more
than 4,500 metres at the Yellai-Igaisk deposit. This is the greatest depth at
which oil has been obtained in Western Siberia.

Although the USSR no longer issues annual data for gas reserves, it is known that
Western Siberia had 27,000 bn cu m at the beginning of 1981. As well as Urengoi,
with 7,500 bn cu m, there are five deposits with more than 1,000 bn cu m of
reserves including Yamburg (4,440 bn), Zapolyarnyi (2,670 bn), Medvezhe (1,550 bn),
and Kharasavei and Bovanenkovsk with at least 1,000 bn cu m each. The Tenth FYP

target for new reserves was fulfilled 18 months early, with 7,000 bn cu m being found in 3½ years, and 10,500 bn cu m were found over 1976-1980. Most of this lies in the area of the Urengoi deposit which has still not been fully explored.

The special problems of Salym and Russkoe

Salym and Russkoe are two of the largest fields to have been found in Western Siberia, but their development is proceeding slowly because of the technical problems they present.

Salym was discovered in 1963 in the Bazhenovskii shale formations which stretch for 1,000 kms from north to south and several hundred from east to west. Some Western sources speculate that its recoverable reserves may equal Samotlor's initial 2,000 mn tons[1] and the total in-place reserves of the Bazhenovskii formation have been estimated at as much as 620 billion tons, although the recovery rate is unlikely to exceed 20%.[2] While some 15 deposits have been found in these shales, Salym is the most significant. Lying 100 kms west of Surgut, it has seven payzones, although only one of these, at a depth of 2,800 to 2,850 metres, is of practical significance.

The deposit has anomalously high pressures of 421 ats, compared with normal pressures for depths of 2,800 metres of 289-185 ats. But the pressure falls quickly when wells begin producing, and well-yields decline. Salym began trial production in August 1974, and although three exploration wells gave 80,000 tons of oil in the first year, the pressure fell from 422 to 341 ats. Consequently, well-yields fell very rapidly; at Well 28, when a 6 mm choke was employed, the yield fell from 300 cu m/day in April 1974 to 200 in August 1974. But when a 4 mm choke was used, it fell from 102 cu m/day to 75 in April 1975, and then stabilised at this level.

In 1976, the deposit produced 1.4 mn tons from 25 wells. Research into its geological structure is continuing, and on October 4th, 1979, a controlled nuclear explosion was carried out in an attempt to fracture the rock and stimulate petroleum migration. Evidence has recently appeared that the Russians intend to develop Salym in a big way during the 1980s. Ministry of Oil officials have confirmed the commercial prospects of some of the Bazhenovskii formations deposits as a result of the experimental exploitation of wells in the Salym area[3], and the Transgidromekhanizatsiya trest has been building artificial islands for the drilling of more wells.

Russkoe deposit is situated along the Russkoe river, 680 km north-north-east of Samotlor, and is the largest heavy oil deposit discovered so far in Western Siberia. It was discovered in 1968 and eleven rigs were engaged in exploratory drilling in 1977. Experimental work was being carried out in 1978 by the Zapolyaryeneft directorate of Nizhnevartovskneftegaz, but it is unlikely that production will take place on a significant scale before the mid 1980s.

---

1 Oil and Gas Journal 5/3/79 2 Petrostudies, December 1980 3 Mingareev et al, "Tekhnicheskii progress v neftyanoi promyshlennosti, 1981 p. 7.

Much of the Soviet exploration effort is still taking place in the aging and heavily worked regions of European Russia such as the Tatar and Bashkir republics and Kuibyshev oblast. However, it is not expected to find large new oilfields in these areas. Apart from Western Siberia, only the Komi republic, the Caspian area and Eastern Siberia hold out the promise of the large new finds.

Komi Republic

The search for oil and gas is carried out by Pechorageofizika, which has 23 field parties operating in 1982 compared with 14 in 1975. They are closely followed by wildcatters of Ukhtaneftegazgeologiya, which was responsible for exploratory drilling at 36 of the 42 areas examined during 1976-1980 (Komineft and Komigaz-prom were also involved in exploration work) and has prepared 12 more structures with an area of 1,150 sq kms for drilling during the Eleventh FYP. At Kharyaginsk, the deepest polar well has been drilled to a depth of 4,300 metres and the search for oil has continued in the Kozhvinsk, Izhma-Pechora, Kilvinsk, southern Khoreiversk and Varandei-Adzinsk areas.

Exploration for gas has been largely confined to looking for deeper payzones at the Vuktyl gas condensate deposit. Four wells have been drilled below 4,750 metres, and a new extension of the shallower payzone has been found. Exploration of the Pechora-Kozhvinsk, Shapkino-Yur'yakhinsk, Laisk, Denisovsk and Verkhne Pechora zones is continuing. Some 75 bn cu m of new gas reserves must be found according to the Eleventh FYP, and exploratory work will extend northwards into the Arctic Ocean. A new Arktikmorneftegazrazvedka Association has been set up at Murmansk, and sand islands are being created by Transgidromekhanizatsiya. The Vyborg Shipyard has started building semi-submersibles similar to the "Shelf" series being built at Astrakhan for the Caspian, and the Russians expect to take possession of three dynamically self-positioning drilling ships from Finland's Rauma Repola shipyard before the end of 1982. The first arrived at Murmansk in March.

Kazakhstan

In the past, exploration work (carried out by Kazneftegazrazvedka, based at Gurev) has been undertaken mostly on the Mangyshlak Peninsula. The two giant fields of Uzen, with 500 to 700 mn tons of oil according to Western estimates, and Zhetybai with 150 mn tons have been producing for over 20 years, but output has been declining due to the high viscosity and paraffin content of the oil. Recent exploration has focussed on wells of more than 6,000 metres in an attempt to find less viscous oil at Mangyshlak, and very shallow wells in the heavy oil region of the Buzachi Peninsula.

Work by Embaneft on the aging Emba field of Gurev and Aktyubinsk oblast has resulted in the discovery of the Tengiz deposit, and this is said to confirm the good potential of the region for more finds. The exploration of subsalt structures has revealed the Dosmukhambetovsk and Vostchno Kokarna deposits and new pools at the aging Dossor field. The finds in southern Aktyubinsk oblast will begin producing during the Eleventh FYP period.

The Caspian Sea

Exploration work on the Caspian Sea's Apsheron Sill proved disappointing during the first three years of the last FYP period, with the plan for new reserves being underfulfilled by 82 mn tons. However, the discovery of the major

"28th April" deposit in 1979 may have reversed this trend, and with new jack-ups and semi-submersibles operating, the Russians expect enough new oil to be found during the next few years to enable off-shore production to increase.

At least 18 structures remain to be drilled, with 11 close to the Azerbaidzhan shore, two to the north of "28th April" and three to its east (Ushakova, Kaverochkina and "26 Bakinskikh Kommissarov"). On the Turkmen side, Promezhutochnaya and Livanov-Zapadnyi remain to be drilled. Appraisal drilling at Zapadnyi Erdekli is continuing, with three wells finding oil at 3,200 metres.

In 1978, 31 exploration and appraisal wells were drilled in the Caspian; this figure rose to 42 in 1979, and it can be expected to rise throughout the 1980s , with the volume of exploratory drilling set to rise from 350,000 metres in 1980 to 1 mn metres in 1990.

Azerbaidzhan On-shore

Apart from the Kura-Iori interfluve, where large volumes of oil are believed to exist at depths of more than 5,000 metres and the first well gave pay of 40 tons a day in 1981, Azerbaidzhan is not likely to be the scene of major new oil discoveries, although 17 small oil and gas deposits were found during 1976-1980. Its interest for geologists lies in the drilling of the Saatli well which is expected to reach an eventual depth of 15,000 metres. A fully automated rig brings up core samples at 50 metre intervals, with complex geophysical work carried out every 250 metres. Drilling began in 1977, and by September 1981 it had reached a depth of 7,500 metres. While the first 7,500 metres took $4\frac{1}{4}$ years to drill, the second 7,500 is planned to take 5 years with the well completed by the end of 1986. A Uralmash-15000 rig with a payload of 400 tons is being used.

Astrakhan

Although gas shows were obtained in Astrakhan oblast as long ago as 1952, the first important finds were made when a condensate well gave 0.6 mn cu m a day from 4,000 metres in 1976, and a second well gave a similar yield in the eastern part of the oblast in 1977. It is now thought that the deposit covers 30,000 sq km, stretching from the centre of the Kalmits republic in the west to the Kazakh steppe in the east. This makes it the largest gas deposit in the European USSR, and its reserves have been tentatively estimated at 6,000 bn cu m. It is also believed that the deposit has many pay-zones, with the deepest at 7,000 metres likely to contain oil. The Ministry of Geology is carrying out a programme involving the drilling of 14 exploratory wells to 4,200 metres.

The comparatively great depth of the deposit means that it will be exploited according to a long time profile, with a small number of wells operating over a long period. Current plans foresee the drilling of 50 development wells giving 6 bn cu m a year with the introduction of further stages raising output to 18 bn cu m a year during the 1990s. However the plan to complete the first stage of exploration work by 1979 has still not been fulfilled because the poorly-equipped Astrakhan Oil and Gas Exploration Expedition is able to drill only 10-12,000 metres a year instead of the necessary 40,000. Accordingly, the plan to begin exploitation in 1985 is unlikely to be met.

Turkmen Republic

The Turkmenneft Association is drilling exploratory wells for oil at 30 areas. Most of the new oil reserves are being found in the western part of the republic, principally at the existing Cheleken, Kotur Tepe and Barsa Gelmes deposits. The Okarem area is said to be promising, although comparatively little work

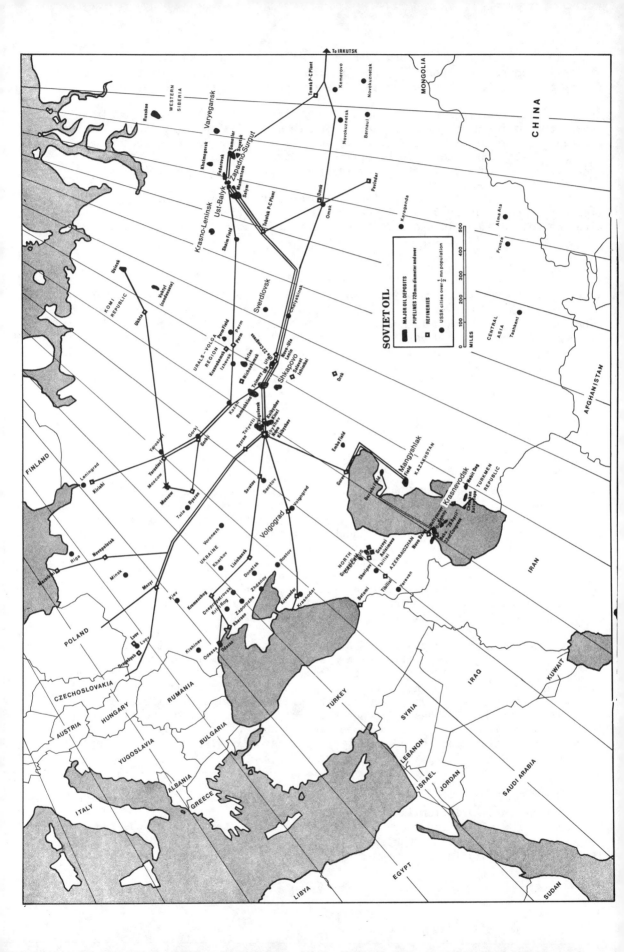

SOVIET OIL

MAJOR OIL DEPOSITS
PIPELINES 720mm diameter and over
REFINERIES
USSR cities over ½mn population

MILES
0   100   200   300   400   500

To IRKUTSK

CHINA

MONGOLIA

Tomsk P-C Plant
Kemerovo
Novokuznetsk
Novokuznetsk
Barnaul

Pavlodar

Karaganda

Alma Ata

Frunze

CENTRAL
ASIA

Tashkent

AFGHANISTAN

IRAN

WESTERN
SIBERIA

Russkoe

Varyegansk

Khalmegansk
Federovsk
Samotlor
Sugutsk
Magistovo
Salym

Zapadno-Surgut
Ust-Balyk

Krasno-Leninsk

Shaim Field

Omsk
Tobolsk P-C Plant
Omsk

Sverdlovsk

Chelyabinsk

KOMI
REPUBLIC

Ukhta

Usinsk

Vuktyl
(condensate)

URALS - VOLGA
REGION

Pena Field
Perm
Krasnokamsk
Izhevsk
Bizhenikamsk
Alan

Nova-ufa
Ufa
Salavat
Ishimbai

Orsk

Tolyatti Syzran
Kazan
Romashkino
Kuibyshev
Dyrshino
Kinel
Kalipahov

Shkapovo

Emba Field

Mangyshlak
Field

KAZAKHSTAN

Krasnovodsk

Nebit Dag

Okharem
Kotluken

Guryev

TURKMEN
REPUBLIC

Ukhinsk

Yaroslavl
Gorki

Yaroslavl
Moscow

Moscow

Tula Ryazan

Stratov

Saratov

Volgograd

Volgograd

FINLAND

Leningrad

Kirishi

Novopolotsk

Riga

Minsk

Mozyr

Kiev

Voronezh
Kharkov
Lisichansk

Donetsk
Zhdanov

Rostov

Krasnodar

UKRAINE

Kremenchug
Dnepropetrovsk
Krivi Rog
Zaporozhe
Kherson

Kishinev
Odessa

Lvov
Lvov

Drogobych

POLAND

CZECHOSLOVAKIA

AUSTRIA
HUNGARY

YUGOSLAVIA

RUMANIA

BULGARIA

ALBANIA

GREECE

ITALY

Krasnodar

NORTH
CAUCASUS
Shaulippyi
Tbilisi
Batumi
Tbilisi
Yerevan

Groznyi

Makhachkala
Gurgany
Artiobagen
Neva Bozi

AZERBAIDZHAN

Baku
Darichensen

TURKEY

CYPRUS

LEBANON
ISRAEL
JORDAN

SYRIA

IRAQ

KUWAIT

SAUDI ARABIA

EGYPT

LIBYA

SUDAN

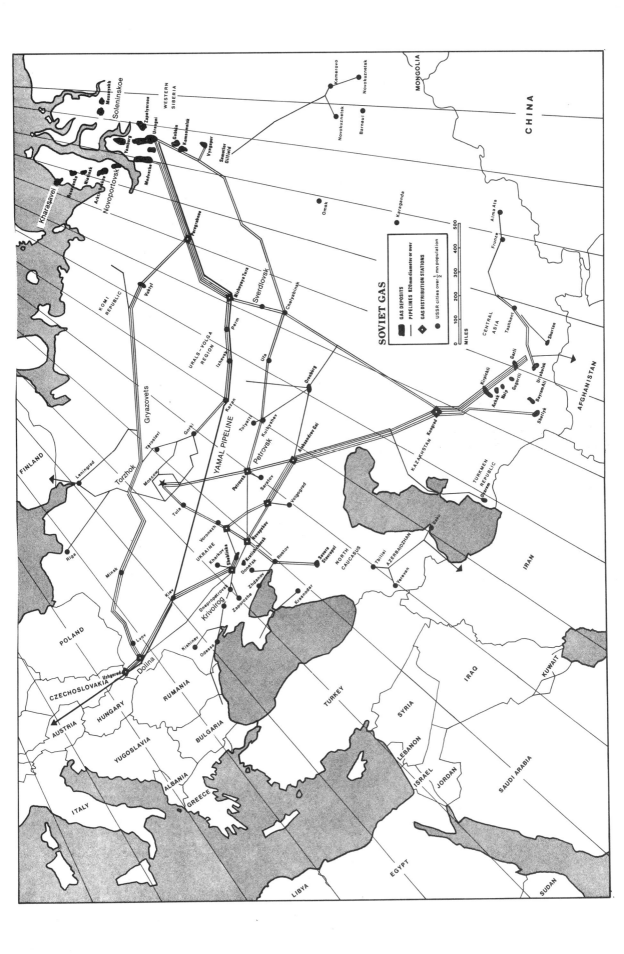

SOVIET GAS

GAS DEPOSITS
PIPELINES 820 mm diameter or over
GAS DISTRIBUTION STATIONS
USSR cities over ½ mn population

MILES
0   100   200   300   400   500

CHINA

MONGOLIA

Kemerovo
Novokuznetsk
Novokuznetsk
Barnaul
Novokuznetsk

WESTERN SIBERIA

Soleninskoe
Messyakha
Zapolyarnoe
Urengoi
Gubkin
Komsomolsk
Vyngapur
Samotlor Oilfield

Tambeg
Medvezhe

Kharasavei
Novoportovsk
Arkhangelsk
Medvezhe
Molotovsk

Omsk
Karaganda

KOMI REPUBLIC
Vuktyl
Pervomaisk

CENTRAL ASIA
Alma Ata
Frunze
Tashkent
Chardzhou
Gazli
Kungrad
Kiyakti
Achak
Naip
Gugurtli
Bayramali
Shatlyk

FINLAND
Leningrad
Riga
Minsk

Torzhok
Gryazovets
Yaroslavl
Gorki
Moscow
Tula

URALS-VOLGA REGION
Nizhnyaya Tura
Perm
Sverdlovsk
Chelyabinsk

YAMAL PIPELINE
Izhevsk
Kazan
Toyatti
Kuibyshev
Ufa
Orenburg

Petrovsk
Penza
Saratov
Volgograd
Aleksandrov Gai
Kungrad

KAZAKHSTAN

TURKMEN REPUBLIC

AFGHANISTAN

IRAN

POLAND
Lvov
Uzhgorod
Dolina

CZECHOSLOVAKIA
AUSTRIA
HUNGARY
RUMANIA
YUGOSLAVIA
ALBANIA
BULGARIA
GREECE

UKRAINE
Kharkov
Sheblinka
Donetsk
Rostov
Severo Stavropol
NORTH CAUCASUS
Tbilisi
AZERBAIDZHAN
Yerevan
Baku

Voronezh
Novopskov
Krasnodar
Zhdanov
Zaporozhe
Dnepropetrovsk
Krivolrog
Kishinev
Odessa
Kiev

TURKEY
SYRIA
LEBANON
ISRAEL
JORDAN
IRAQ
KUWAIT
SAUDI ARABIA

ITALY
EGYPT
LIBYA
SUDAN

has been carried out. In 1976-1980, 6 oil and gas deposits were discovered, including the important Erkekli find, and results were sufficiently encouraging for the Eleventh FYP to call for 760,000 metres of exploratory wells to be drilled over 1981-1985.

Eastern Siberia

By 1982, 42 oil, gas and gas condensate deposits had been discovered in the Eastern Siberian regions of Krasnoyarsk krai, Irkutsk oblast and the Yakutsk republic, but only the Messoyakh and Solenoe gas deposits, serving Norilsk, and the Mastakh and Ust Vilyuisk gas deposits, serving Yakutsk, are being worked.

The most studied areas are Vilyui, Nepsko-Botuobinsk and Angara-Lena. In the central part of the Siberian plateau, i.e. the Nizhnetunguska, Severo-Aldansk and Priverkhoyansk regions, the volume of drilling varies between 0.001 and 0.08 metres per sq km, but in the central and northern parts of the Tunguska syncline, not one well has been sunk. The drilling density is only 6 per cent that of the USSR as a whole. The level of geophysical study is very low; half the area has not been covered at all, and detailed seismic work has not been carried out on more than 20 per cent.

In the Vilyui zone, more than 30 gas and gas condensate deposits have been found, including two (Taas-Yuryakh and Verkhne-Lindensk) in 1981, but so far a total of only 5 bn cu m of gas has been produced from two deposits for local needs. The region's importance lies in its gas condensate reserves, but it seems that insuf-ficient effort is being made to develop them. Mastakh is now flaring 30 cu m of condensate a day, equivalent to 6 tons of petrol and 8 tons of boiler fuel. This is because the production of condensate has not been catered for by the local Yakut planning organisations. The Yakut Gosplan has "solved" the problem by transferring storage tanks belonging to Lenaneftegazgeologiya from the Kysyl Sur and Khatyryk-Khomo areas. Mastakh will produce 7,000 tons of condensate in 1982, but its rational utilisation has still not been organised. Exploratory work in the Vilyui region is now being concentrated on the Khapchagai region, where several new gas deposits have been found, and on proving up the reserves of the Sredne-Tyungsk deposit.

In the Botuobinsk region, a well drilled by the Srednelensk expedition of Lenaneftegazgeologiya Association has given a commercial flow of oil from 1,900 metres. This was the first oil flow in the Yakut republic, and oil has also been found recently at one of Yakutia's largest gas deposits, Srednebotuomsk.

The Yeniseineftegazgeologiya Association is responsible for exploring the Eastern Siberia platform. Flows of oil, gas and condensate have been obtained by the Turukhansk expedition in the Yenisei valley in the region between the Kureisk and Letninsk tributaries, and a gas deposit has been found in the Khobeisk area by the Nizhneyeniseisk expedition. The first well yeilds 1½ mn cu m from 2,046 metres, suggesting that it may be a very large find.

In the south west of the Siberian platform on the Kamovskii arch, several shows of oil have been obtained, and the Evenki expedition has found oil at Vanavara and in the Nizhnyaya Tunguska region, principally along the Kochechym tributary. In the Danilovo area, 200 kms south of Yerbogachen on the Nizhnyaya Tunguska, appreciable reserves of gas have been discovered, and the first oil was found in November 1980 when a well yielded 100 tons a day. The most important oil find has been that of Kuyumba on the Podkammenaya Tunguska river. where commercial flows of oil were discovered at a depth of 2,340 metres in Paleozoic reefs.

The search for oil and gas is gathering pace in the Taimyr Peninsula and the northern coast of Krasnoyarsk krai. Commercial gas flows have been obtained at the Balakhinsk site on the northern shore of Khatanga Gulf, and prospects are believed to be good for oil on its southern shore in the Novyi region. Further east, the Khatanga Oil Exploration Expedition has found oil shows in the valley of the Anabar river. Seismic field parties are now working on the Taimyr Peninsula, especially in the Portnyagino region to the north of the Balakhinsk discovery. The existing gas deposits of Solonoe and Messoyakh, now giving 3 bn cu m a year, are being developed further, with the yield of Solenoe planned to double to 0.6 bn cu m a year. Two new gas deposits, Severo-Solenoe and Pelyatkinsk, are being tested.

In the Far Eastern region, gas has been discovered on Kamchatka; the first deposit was discovered in November 1981 on the banks of the River Kshuk. In the extreme north-east of the USSR, the Chukotka expedition has found gas and oil in the Anadyr Depression.

OIL EXPLORATION TECHNOLOGY

The USSR widely employs the reflective method of seismic surveying, and although the refractive method is mentioned occasionally, particularly in relation to the need to study paleozoic strata at great depths in Western Siberia, there are no indications that it has been used.

Efforts have been concentrated on developing modifications to reflective profiling. The greatest limitations are imposed by the inadequate technical level of data-processing centres and a shortage of geophones and experienced crews. These limitations are caused partly by the rapid increase in the volumes of seismic work, and partly by the adoption of Common Depth Point stacking in most regions. For organisations using CDP, the volume of drilling work necessary for shots has grown three to four times.

The shortage of basic and auxiliary equipment has led to surveying being limited to the most promising areas. However, a plan has been drawn up under which the USSR is divided into four types of region, each having similar tectonic structures and geological history. These are:

1. The Russian and East Siberian platforms.
2. Western Siberia, Central Asia and the Caucasus foothills.
3. Western Turkmenia, Azerbaidzhan, Sakhalin and the Carpathian Depression.
4. The Pripyat, Donets Dnepr and Caspian depressions.

These four types of region embrace 90 per cent of the prospective oil and gas bearing area of the Soviet Union. The second type includes Western Siberia where upper Meso-Cainozoic strata provide favourable conditions for seismic profiling with good reflecting horizons and easily mapped, clearly expressed structures. The lower stages are more complex, and their characteristics are studied sufficiently.

Types 2 and 3 are most suitable for seismic work. Their complex lower stages are called "group b" to distinguish them from the upper stages, known as "group a". For "group a" of types 2, 3 and 4, reflective methods using CDP with a low multiplicity factor give satisfactory results, which can be improved by increasing the multiples of soundings. For "group b" structures, reflective methods, even with CDP stacking, do not always give good results.

The best results are obtained by the reflective method in its conventional form for "Type 2 group a" regions and its highest productivity, with an average speed of 97.5 km a month, is achieved in studying these structures.

For "group a" structures at comparatively shallow depths of 2,500 to 3,000 metres, it is often possible to use reflective methods in their usual form. However, intensive prospecting (with density of 1.7 to 2.4 km per sq km) in many regions is caused by the need to examine small structures and non-structural traps of the lithological-stratigraphic type. This requires more accurate seismic profiling, which can be obtained by a three to six-multiple CDP modification.

"Group b" structures include transitional complexes of young platforms, the deep-lying Liocene-Miocene strata of Cainozoic geo-synclinical zones and sub-salt strata. These are neglected, because their investigation by reflective methods with 6-multiple CDP in combination with other geophysical methods does not give satisfactory results. Even CDP in multiples of 12 to 24, with a productivity of only 15 to 23.5 km/month, does not always give good results, and must be accompanied by parametric drilling and vertical seismic profiling.

Sub-salt strata in type 4 regions are most difficult to study. They require 30-multiple CDP stacking with parametric drilling and vertical seismic profiling. In Western Siberia, CDP stacking was used for the first time in 1973, at Ombinsk and Ob-Yugansk in the Middle Ob and Kurgloi in the Shaim field. It is said to have given clear, continuous readings of profiles in Jurassic structures to depths of 3,300 metres, thereby raising a question mark over the assertion by an American journalist that "the best Soviet gear probes down to 2,300 metres."[1]

The Russians do not dispute that their seismic data processing technology can be improved. American expert Rigassi has noted that "during 1975-1977, drilling on seismic prospects had shown 30 per cent of the postulated structures to be unclosed. The figure increases to 75 per cent in certain structurally complex areas, and in dealing with deep prospects."[2]

They do claim, however, that they are making considerable headway in improving their technology. This has primarily involved better computer technology for data processing, the more extensive use of Common Depth Point Stacking (stacking with a multiple of 6 and more accounts for 65% of all Soviet seismic work at the present time), the greater use of vibration devices (which obviates the need to drill holes for seismic shots) and improvements in data storage equipment.[3]

Perhaps the most rapid improvements are taking place in the sphere of computer technology. Not so long ago, most seismic data collection work was accomplished by analogue methods, and a recent study by the US Congress Office of Technology Assessment asserts that "in the USSR, roughly 40% of collection work is still done by traditional analogue methods". The Russians, on the other hand, claim that "an important achievement during the Tenth Five Year Plan period (i.e. 1976-1980) was the full transfer to digital data processing". A computer centre equipped with 2nd or 3rd generation computers with specialised software has been established for each geophysical trust, and the plan to create 40 of these centres by 1980 was fulfilled in spite of the continuing shortage of skilled operators. To aid the storage of seismic data, the "Kvant" device for the transfer of data from a computer to microfilm has been designed by the Neftepribor plant and its serial production has begun.

1 _Time_ magazine 23/6/80 2 World Oil 15/8/79 3 Economist Intelligence Unit: Quarterly Energy Review, Soviet Union and Eastern Europe. 1982 No. 1

During the last five years, standard programme packages have been elaborated, notably by the Central Geophysical Office of the Ministry of Oil. The Russians claim that some of these, like the STsOGT-2 system, incorporate all the most widely-used algorithms.

Improvements in the technology of field systems have been largely due to the efforts of the Special Design Bureau of Seismic Engineering in Saratov. Among the many instruments it has designed, special importance is attached to the Potok series of input and output machines, and devices enabling drums, discs and magnetic stores designed for ES type computers to work with BESM-4 and M-222 computers. The Bureau is most proud of its digital seismic field stations. These include the "Volzhanka" which, according to the Bureau, meets world standards for capability and reliability, and which carries out recording in the international format "seg B" permitting a direct link between the seismograms and the computer, thereby avoiding the need for intermediate equipment. It has also built the "Progress-2" which processes explosive seismic waves, and the "Progress-3" to record vibration waves. The "Impuls" Scientific Production Association of Severodonetsk is also charged with building field stations. Its "PS-2000" data processing complex has been tested in Western Siberia and the Urals and is said to work rapidly and efficiently.

The Design Bureau of Seismic Technology in Gomel is responsible for developing seismic wave emission apparatus. Its more recent designs include the SI-32 and SI-64 gas detonation impulse devices which work automatically by radio control from the seismic station. New percussion, electric spark and gas detonation devices have been elaborated as well as vibration instruments such as the "Vibrolokator" which permits the study of profiles up to 10,000 metres deep.

Several organisations in Western Siberia have joined forces to produce a portable remote-control field recording unit, the "Taiga", for remote regions. It employs a seismic tape recorder using simple tape-extending mechanisms with a range of more than 50 dB, and a noise-stabilising system of radio remote control. The "Taiga" is now widely used in Siberia and the Far East, and is based on cross-country vehicles or helicopters.

The production of better geophone cables has received particular attention. The Office of Technology Assessment study says that "American experts who have examined Soviet geophone cables have found that they introduce extraneous noise into the data". Since then, the Tatneftegeofizika trust in Kazan has built a workshop for the serial production of an improved 27-ply geophone cable.

Finally, considerable progress has been achieved in marine geophysical research. The semi-automatic "Grad" system involves an on-board analogue-digital complex based on a Minsk-32 computer which processes seismic data obtained with CDP stacking. An improved version based on an ES-1010 computer has been designed by the Yuzhmorgeo organisation jointly with Hungarian specialists, and is now being tested.

However, it will be several years before these technological advances begin to show up in better exploration work. The bottleneck is now one of a shortage of skilled operators.

Geophysical work in Western Siberia is being assisted by the Western Siberian Research Institute of the Geology of Oil and Gas, headed by Academician I. Nesterov. It is currently drawing up maps of the Bazhenovskii oil shales, and designing the new technology required for the anomalously high pressures. During the last five years, the institute has been working on the elaboration of mathematical models, which, with the use of geological, tectonic and hydrological data processed by the ES-1020 computer, can calculate the optimal number and location of exploratory wells. This research has been timely, because Nesterov says that in recent years the effectiveness of deep exploratory drilling has fallen and costs have risen. This is due to exploration work moving northwards into remote regions with complex geology.

The success of their work obviously depends on improvements in seismic technology, and it is significant that the organisation responsible for this, the Special Design Bureau of Seismic Engineering in the Volga city of Saratov, continuously wins praise for its work in this respect.

LABOUR

In 1975, the Ministry of Oil employed 752,200 workers, and this number had been growing at an average annual rate of 5.2 per cent since 1966 when 479,660 workers were employed. Since 1975, the growth rate of oil production has matched that of labour productivity, suggesting that employment has stabilised at about 750,000. Labour productivity in the gas industry is very high, and consequently it employs only a fraction of the number of workers employed by the Ministry of Oil. The number of gas workers has been growing slightly. During 1976-1980, output by the gas industry rose by 50.5% while labour productivity rose by 44 per cent, suggesting an increase in the number of workers of 4.6 per cent.

As the Soviet oil industry has become more technologically sophisticated, the share of workers with a higher or middle specialist education has grown, from 11.8 per cent in 1966 to 20.5 per cent in 1975. It is interesting to note that within this group, the share of workers with higher education has declined slightly from 41 per cent to 39 per cent.

Most of the specialists are concentrated in the older producing regions where the problem is to stem the decline in output. From an estimated 175,000 specialists in 1977, only 18,805 were working for Glavtyumenneftegaz which produced 210 mn tons of oil, while nearly 9,000 (half with a higher education) worked in Azerbaidzhan, producing 16 mn tons of oil.

The oil workers with a higher education are trained in seven vuzy (technical colleges) and six university faculties. A middle education of oil production can be obtained in 21 tekhnikums under the jurisdiction of the Ministry of Oil, and which are situated in all the oil-producing regions. There are also numerous other institutes where an education in particular aspects of the oil industry can be acquired. A typical example is the Tyumen proftekhuchilishcha No. 7 which trains drillers, rig assemblers and exploration specialists.

By far the most important educational institution is the Moscow Institute of Petro-chemical and Gas Industries. One in six of all specialists in the oil, gas and petrochemical sectors (40,000 in all) have graduated from the institute, and it currently has 11,000 Soviet students as well as hundreds from overseas.

Research institutes are assuming a growing importance. The number of workers engaged in research has grown from 11,106 in 1964 to 24,581 in 1976, including 9,375 scientists. The number of research workers has grown particularly rapidly during the 1970s.

The Soviet oil and gas industries are organised into associations (obedinenii) normally on the basis of one per oblast or republic. In the case of Tyumen oblast, the oil association, Glavtyumenneftegaz, is subdivided into four other associations (see Chapter 2). Each association consists of a number of directorates (upravienii), each specialising in a particular activity such as extraction (NGDU), drilling (UBR), transport, well repair, etc. In turn, each directorate is staffed by workers organised into crews (brigady).

At the end of 1977, the Ministry of Oil had 5,400 crews, including 1,246 for drilling, 339 for rig assembly, 1,079 for well repair, etc. Each crew has, on average, 140 members. The 1,246 drilling crews were employed by 115 UBRs, which belonged to 26 associations. In 1978, the number of drilling crews declined to 1,199 but rose again in 1979 to 1,231.

Like all Soviet workers, drillers are paid a bonus for fulfilling their targets, and a further bonus if they overfulfil them. It is commonly believed in the West that a major problem stems from drillers being paid bonuses on the basis of a plan target consisting of the number of metres drilled. They would obviously be tempted to fulfil their targets by drilling a large number of shallow dry wells, and there is evidence that this has happened in a number of cases.

In 1977, the Ministry of Oil introduced a new system of incentives. Bonuses are now paid to all the workers of an association on the basis of the volume of oil produced by the association, the so-called "potonnaya stavka". The problem now consists of how to distribute the money in the incentive fund between all the association's directorates, some of which do not enjoy a direct link between their work and the volume of oil produced.

The Tatneft association (Tatar republic) has elaborated its own system. The size of the incentive fund for directorates of a particular type (e.g. UBRs) is determined by the wages fund of the directorate multiplied by a standard rate, and then multiplied by a "correction factor", the size of which depends on the type of directorate. For the group of drilling directorates, the correction factor is 1.3; this compares with a factor of 1.15 for the group of oil-producing directorates and 0.5 for the group of transport directorates, and testifies to the importance of drilling work to the success of the association as a whole.

Within the group of drilling directorates, each directorate is allotted an incentive fund, the size of which depends on the results of another complicated formula. This formula incorporates three indices:

   i.   the labour productivity of the drillers;

  ii.   the increase in reserves of oil and gas revealed by drilling; and

 iii.   an index determined by the quantity of wells which have been completed and are producing oil multiplied by the period of their operation from the moment of handing over to the extraction directorate to the end of the control period; this index is measured in well-days.

The last two factors effectively bar the inclusion of dry wells in the basis on which incentives are calculated, and discourage the drilling of such wells for the purposes of earning bonuses. In addition, the third index is designed to encourage the drillers to reduce the duration of the well-drilling process. The shorter the time taken for construction of a well, the greater the number of days it is producing oil until the end of the control period, and the greater the number of well-days chalked up by the drilling enterprise.

The whole procedure is designed to speed up the creation of oil-producing capacity, and the drillers, like everyone else, only get their bonuses if the association fulfills its plan target for the production of oil. The first results of the Tatneft experiment are encouraging. In 1977, the average well-construction time fell by 8.3 per cent and, for development drilling alone, it fell from 53.1 days in 1976 to 46.3. Other associations are adopting similar systems of bonus distribution.

CHAPTER 6

The Technology of the Hydrocarbon Industry

Backward technology has been identified by all Western writers as being the weak spot of the Soviet oil and gas industry. Most writers, however, have made this point and then assumed that this will prevent oil and gas production targets from being fulfilled.

It can be argued, on the other hand, that the potential for improvement in technology is one of the USSR's greatest assets. If it can keep up the continual improvements in equipment productivity that have been registered during the 1970s, then it should be able to produce increasing volumes of oil with the same number of workers (it has already been mentioned that the labour force has stabilised at 750,000) and the same number of rigs and other equipment. In fact, the last few years have seen a decline in the size of the rig park and a substantial decline in the number of drilling bits and turbo-drill sections manufactured. This has been made possible by huge increases in productivity.

The production of oil and gas equipment, with the exception of drilling rigs, is undertaken by the Soyuzneftemash Association belonging to the Ministry of Chemical Engineering. This organisation has 30 factories, including 13 in Azerbaidzhan which has traditionally been a centre of the petroleum engineering sector, and which produces 70 per cent of Soviet oil and gas field equipment, mainly pumping units, pumps, well-head valves, well-testers etc. Most of the other factories belonging to Soyuzneftemash are administered by Volgogradneftemash in Volgograd and Saratovneftemash in Saratov. It is planned to set up an engineering base in Tyumen and this, with the reconstruction of the Azerbaidzhan plants, is one of the principal features of the Eleventh FYP for petroleum engineering.

The production of oil apparatus has been growing steadily, particularly since 1970, although the production of some items has been declining because new models are being produced with much higher productivities than those they replace.

Table 21

	1950	1960	1970	1975	1979	1980
Oil apparatus ('000 tons)	47.9	93.0	126.6	135.0	188	184
Pumps ('000)	65.7	81.8	77.0	85.1	95.0	91.6
Turbo-drills	978	6,222	6,562	9,780	8,976	9,270
Electro-drills	4	107	115	97	21	0

Source: Naradnoe khozyaistvo 1980. After 1970, "oil apparatus" excludes anti-air-pollution equipment, which weighed 36,000 tons in 1975.

The biggest plant producing equipment for oil and gas fields is the Bakinskii Rabochii plant in Baku. It is also one of the fastest growing plants, and it is expected to increase its production level by 80 per cent over the Tenth FYP period. The other large Baku plants are the Leitenant Shmidt, Lenin, Kishlinsk, Sardarov and Montin. During the 11th FYP, the Leitenant Shmidt plant will account for most of the increased output of equipment from Azerbaidzhan, with its volume of output planned to rise by 67 per cent over 1981-1985.

During the period 1970-1978, the average output of oil per Ministry of Oil rig increased by 64.8 per cent, by 62 per cent per crew and by 46.5 per cent per drilling worker. This was largely due to vastly improved commercial speeds of drilling, particularly by development rigs. These were assisted by the easier drilling conditions of Western Siberia which is accounting for an increasing share of drilling. Drilling in Siberia generally takes place in simple geological conditions and soft rocks such as clays and sand, and is mostly accomplished from artificial islands from which clusters of up to 20 inclined wells are drilled. This enables Siberian drilling speeds to average 3,596 metres a month, compared with 1,550 by the country as a whole, in 1976. Some crews drill over 8,500 metres a month. The crew of G. Levin has drilled over 100,000 metres a year for several years, and since it arrived from Kuibyshev in 1968 has drilled over 1 mn metres.

Table 22

Commercial drilling speeds (metres per rig per month)

	Exploration	Development
1950	209	629
1960	401	993
1970	339	1,120
1975	336	1,485
1980	411 (1979)	1,791

Sources: Naradnoe khozyaistvo 1965, Neftyanoe khozyaistvo 1977 No. 10
         Neftyanoe khozyaistvo 1980 No. 5, Mingareev et al op cit
         Ekonomika neftyanoi promyshlennosti 1981 No. 12

The 1980 commercial speed of 1,791 metres per rig per month compares with the assertion by a Western journalist that "it takes a Soviet team 14 months to dig down to 10,000 feet (3,048 metres)."[1] In fact Soviet drillers in Western Siberia, where an average speed of 3,596 metres a month was achieved in 1976, work faster than American drillers who can only manage a rate of 2,690 metres per rig per month according to Time magazine.

Drilling speeds have increased in spite of growing well depths. The average depth of new development wells has grown from 1,586 metres in 1960 to 1,772 in 1970, 1,994 in 1978 and 1,995 metres in 1979. The average depth of exploration wells has grown from 1,928 metres in 1960 to 2,775 in 1975 and 2,797 in 1978 before falling to 2,770 metres in 1979.

It can be seen from Table 22 that drilling speeds have been accelerating since 1970, and for exploratory drilling, since 1976. It is difficult to apportion the credit for this improvement among such factors as easier drilling conditions, more cluster drilling, better organisation or better technology of rigs, bits, turbo-drills, drilling fluids, etc. But there is no doubt that the steady improvement in the technical level of the rig park has played a role.

The number of rigs operating in the USSR has actually declined since 1970 from 2,083, including 1,124 operated by the Ministry of Oil, to 1,915 in 1978, including 1,013 belonging to the Ministry of Oil, 336 to the Ministry of Gas and 566 to the Ministry of Geology. In 1979, the number of Oil Ministry rigs rose by 71 to 1,084, including 644 devoted to development drilling and 440 to exploration and appraisal work.

1   Time   23/6/80

During the 1970s, the average size and technical characteristics of Soviet rigs improved greatly as old rigs are scrapped and advanced new ones begin work. Consequently, a declining rig park has been able to drill an increasing metrage of wells.

In 1975, about half of all exploration drilling was carried out by the Ministry of Oil and the rest by the Ministry of Geology. The Ministry of Gas carries out a small amount of exploratory drilling in Western Turkmenia. But in recent years, the volume of exploratory drilling by the Ministry of Oil has been falling (by 35 per cent since 1965) as explorationists have moved out to new areas. In Tyumen, for example, the Ministry of Oil is responsible for only 2.4 per cent of exploration work, 2.2 per cent in Tomsk, 1.9 per cent in the Komi republic, and 1.5 per cent in the Udmurt republic. In 1979, exploratory work accounted for only 16.3 per cent of Ministry of Oil drilling.

Table 23

Drilling for Oil and Gas, mn metres

	Exploration	Development	Total
1950	2.13	.16	4.28
1960	4.02	3.69	7.71
1970	5.15	6.74	11.89
1975	5.42	9.75	15.17

Drilled by the Ministry of Oil only

	Exploration	Development	Total
1975	2.73	8.93	11.66
1976	2.5	9.6	12.1
1977	2.4	10.4	12.8
1978	2.4	11.7	14.1
1979	2.5	13.0	15.5
1980	2.4	15.8	18.2
1981	2.4	19.4	21.8
1985 Plan			35.0

Sources: Narodnoe khozyaistvo 1975, and Neftyanoe khozyaistvo 1977 No. 2, 1978 No. 4, 1979 No. 4, 1980 No. 4
1985 Plan; Ekonomika Neftyanoi Promyshlennosti 1981 No. 6

The USSR has been producing about 500 rigs a year for the last 30 years, although production has been declining recently from 544 in 1975 to 521 in 1980 after falling to only 473 in 1979. Roughly a quarter of these rigs are exported, leaving an average annual addition to the rig park of 350. During the period 1971-1975, drillers received 1,750 new rigs. This would mean that slightly more obsolete rigs are scrapped each year, and that the average length of life of a rig is perhaps six years.

Table 24

Production of Oil Rigs

	1950	1960	1970	1975	1979	1980
Production of rigs	476	447	480	544	473	521

Sources: Narodnoe khozyaistvo

Drilling rigs are built by the Uralmash plant in Sverdlovsk and the Barrikady plant in Volgograd, with a small number of experimental rigs produced in Baku. Uralmash accounts for 70 per cent of annual Soviet output, although its rigs carry out only 40 per cent of drilling at depths greater than 2,500 metres where the Barrikady rigs are more suitable.

The basic Soviet rigs for nearly 30 years have been the 3D and 4E produced by Uralmash. Serial production of these rigs, together with the 5D and 6E, began in 1951-1954, and modifications and improvements have been made on several occasions. In 1965, the Barrikady plant began producing rigs able to drill down to 2,500 metres, so production of the Uralmash 5D and 6E ceased.

Since 1964, Uralmash has been producing rigs of a new design, the 125 BD and 125BE, with payloads of 125 tons. They are designed for wells of depth 3,000 to 3,200 metres, and come in a modular unit complete with a derrick, metal platforms and shelters, and automatic systems for the mechanisation of round-trip-operations. Their production was made possible by the radical reorganisation of the Uralmash plant.

During the Eighth FYP period, 1966-1970, six new rigs were developed: the 10E, the IIDE (which could operate with both diesel and electric drive), the 300E and 300DE (with payloads of 300 tons), the 13D and the ED. Technical experience was gained in the use of diesel hydraulic drive and large capacity, high voltage electric motors, in the creation of devices for the mechanisation and partial automation of round-trip-operations, and in the use of chains and propshafts for transmission rather than belts.

This work continued during the 1970s , when attention was turned to the creation of rigs for different climatic and geological conditions, to increasing drilling speeds by a planned 50-100 per cent, and to creating the unique "Uralmash-15000" for drilling to 15,000 metres in Azerbaidzhan and the Kola Peninsula. This rig stands 60 metres high and has a payload of 500 tons. It holds the record for drilling the world's deepest well, the SG-3 well on the Kola Peninsula, which passed 11,000 metres in 1981 and is planned to reach 15,000 metres. The Saatly well in Azerbaidzhan is also being drilled to 15,000 metres by the same type of rig, but for a series of other 15,000 metre wells to be drilled in the Urals, the Caucasus and Western Siberia, a new rig has been developed by Uralmash. The two prototypes were successfully tested in 1981.

During the last plan period of 1976-1980, the Russians concentrated on building rigs for drilling to 5,000 metres with diesel hydraulic and electric drive. These rigs are as compact as possible to assist assembly and transportation. Work is now proceeding in six directions:

1. The production of new equipment for rotary drilling using existing rigs with some modifications. These new rigs are the 125DG-P, 125E, 160DG-Sh and the 160E.

2. The creation of modular rigs with new cranes and transport equipment, enabling them to be assembled and transported either in large blocks on special heavy duty transporters, or in small blocks on platforms pulled by tracked vehicles, or in very small blocks on conventional all-purpose lorries. Experimental modular rigs of types 125DGU, 125EU, 160DGU and 160EU have successfully passed industrial tests, and serial production is beginning. The suffixes stand for "diesel hydraulic universal" and "electric universal", with the numbers representing the payload. These rigs can be assembled by one all-purpose crew

instead of several different crews of earth movers, carpenters, concrete workers, assemblers etc. They are 60 tons lighter than their predecessors, and can drill to 3,300 metres at speeds 33 per cent faster.

3. The development of rigs able to drill to 6,500 metres. These are the 200DG-4 and the 200E-4, and the Russians claim that they are technologically comparable to the best in the world. They have a high level of automation of labour-intensive processes, and their modular construction makes them easy to transport and assemble. The prototypes have begun industrial testing in various regions, including the Chechen-Ingush republic.

4. The elaboration of designs for fully automatic rigs to drill to 4,000 metres with discrete (the 125A rig) or continuous (125NG) accomplishment of round-trip-operations. The 125A is being designed jointly with a Leningrad research institute of the Ministry of Geology, and is claimed to be the world's first fully automatic rig, with all its mechanisms controlled by computer. Weighing 1,000 tons, it drills at speeds 40-50 per cent faster than conventional rigs with only half the normal crew. The first batch was sent to explorationists in late 1980.

    The 125 NG is intended for very deep wells in complex geological conditions. The continuity of round-trip work is accomplished by the employment of two alternatively working groups of hydraulic mechanisms instead of the usual drawworks and travelling block. An experimental prototype is under construction.

5. Special rigs are being built for jack-up and semi-submersible units in the Caspian and Black Seas (see Chapter 9). These are the 6000PE (PE stands for "floating electric", and 6,000 signifies the depth to which they can drill). They are used both for jack-ups operating in 60 metres of water and semi-submersibles working in 200 metres.

6. The final direction in which Uralmash has been working has been production of rigs specially designed for cluster drilling in Western Siberia. Although nearly all rigs working in Siberia are produced by Barrikady, Uralmash has produced several designs. The first batch of rigs was sent in 1977. They were designed to drill ten wells from a single site and, in 1978, a new design from Uralmash, the 2000-LK, appeared in Tyumen oblast. It was designed to perform multiple drilling to depths of 3,000 metres, and the rigs were specially suited for transport, assembly and operation in marshy locations. In 1978, 60 were delivered, and it was planned to send 70 in 1979.

The latest Uralmash rig for Western Siberia is the 3000-Euk, standing for "electric universal rig for cluster-drilling, able to drill to 3,000 metres". The derrick is mounted on a large block, which, together with pipestands mounted on a pipe setback, can be moved for short distances on the artificial island. It is moved with the aid of special motors incorporated into the block. Some designs envisage the rig being moved by pneumatic action along inflated tubes. The rest of the rig's equipment (slush pumps, energy equipment etc) remain on a permanent site on the island. Serial production of the 3000-EUK has now begun, and the first batch was delivered to Tyumen in May 1980.

Of the other technical changes undertaken by Uralmash in recent years, perhaps the most important is the transition, on all rigs, from slush pumps of 2-cylinder reversible operation with a capacity of 1,000 HP to one-way 3-cylinder pumps with capacities of up to 1,200 HP. Their advantages over the 2-cylinder units are their small size (50-70 per cent shorter), their much smoother delivery and more constant pressure and torque. In addition, the quantity of changeable parts has been reduced by 20-30 per cent.

The Eleventh FYP calls on Uralmash to produce 2,025 rigs over 1981-1985, including 445 for Western Siberia. Serial production of several new models will begin, including a rig designed to drill to 8,000 metres. Its derrick will have a radical new design resembling the power transmission pylons being erected for the 1500 kv Ekibastuz-Tambov line, and the rig will weigh 700 tons, representing a saving of 250 tons per rig over the conventional design. It is expected to drill 33% faster than previous models.

The Barrikady plant has been altering the technical characteristics of its products on much the same lines as the Uralmash plant, while concentrating on the production of more powerful rigs designed to drill to greater depths. The basic Barridaky rig is the BU-75 (payload 75 tons) developed during the 1950s. The plant also produces the BU-50 with a payload of 50 tons for drilling to 2,000 metres, the BU-80, for drilling to 2,500 metres, the BU-100 (3,000 metres), BY-125 (4,000 metres), BU-160 (5,000 metres), BU200 (6,500 metres), BU-250 (8,000 metres) and BU10000 (10,000 metres).

While the BU-50, BU-75 and BU-80 (with either diesel or electric drive) are commonly used in the shallow oilfields of the Tatar and Bashkir republics, the BU-125 and BU-160s are favoured by Siberian oilmen. They have stands of either 24, 36, or 48 metres and mud systems of useful volume 150 and 180 cubic metres. BU-75, BU-80 and BY-125 rigs have been adapted for cluster drilling. The rig is conventionally divided into two parts, mobile and stationary. The mobile part consists of the derrick, rotary table and drawworks, and the stationary part includes the pumps, mud tanks, power source etc. Usually, the mobile part is moved from well to well on rails for a distance of 3 metres, which normally takes 20 to 30 minutes, and the stationary part remains in place until all the wells are drilled, but at Nefteyugansk a unit is working where all the stationary features are joined to the derrick under a special roof built by Uralmash. Another method of moving rigs is to use pneumatic wheels from AN-22 planes, capable of sustaining very heavy loads. The rig can then be moved, not only from one well to another on an island, but between islands if they are connected by a hard-surface road.

The creation of firm foundations for rigs in the swamps of Western Siberia and some other regions has been a difficult and expensive problem. The construction of an artificial island through the replacement of waterlogged peat by sand has been the most popular method, especially in Western Siberia, and although this costs an average 44 roubles per square metre, it is nevertheless cheaper than supporting the rig on tubular driven piles (107 roubles per sq m) or screwed piles (220 roubles). However, a shortage of sand in some places and problems with building the islands in summer have made time rather than cost the principal constraint.

During the last plan period, experiments were carried out using new methods of spreading the weight of the rig on sand embankments strengthened by a timber base. It has been shown in Western Siberia that, for a swamp of depth 1.5 metres, the cost of the artificial island can be lowered to 25 roubles per sq metre, but this method is unreliable for greater swamp depths.

Other experiments have involved maintaining the frozen state of the ground throughout the summer months. Under normal conditions, the ground melts to a depth of only 0.5 metres beneath equipment and in the shade, and up to 1.2 metres in direct sunlight. By keeping the designated area free of snow in winter, the ground freezes more rapidly, and is then covered with an insulated flooring and a timber base, which effectively prevents it from melting during the summer months. This frozen state can be maintained practically indefinitely.

The Eleventh FYP expects the volume of drilling by the Ministry of Oil to more than double over 1981-1985 to 35 mn metres a year, with drilling in Western Siberia growing from 7.9 mn metres in 1980 to 18 mn in 1985. Although rig productivity is expected to continue to grow rapidly, it is recognised that the decline in the size of the rig park must be reversed, and the Ministry of Heavy Engineering (Mintyazhmash) has drawn up plans for Uralmash and Barrikady to step up their rig production. It is likely that by 1985, the USSR will be producing more than 650 rigs a year, and the rig park will have grown to over 3,000.

DRILLING BITS

The improvement in Soviet drilling bits can be measured in terms of the metrage of drilling per bit. It can be seen from Table 25 that bit productivity has grown by 10.8 per cent a year over 1970-1979 for development drilling, and 4.7 per cent a year for exploratory drilling. While the 1980 plan target for exploratory drilling bit runs was not fulfilled, that for development drilling bits was substantially overfulfilled.

Table 25

Average bit runs
(metres per bit)

	1970	1975	1979	1980	1985 Plan
Development drilling	33.7	54.2	84.3	88.3	198.5
Exploratory drilling	19.8	26.6	29.8	30.2	

Sources: Ekonomika Neftyanoi Promyshlennosti 1977 No. 10
Neftyanoe Khozyaistvo 1980 No. 5, Mingareev op cit, Trud 6/680

Bit productivity in Western Siberia is far greater than for the country as a whole because wells are drilled into clay and sand, and it has even been suggested that the conditions are so easy that a well should be drilled with only one bit instead of eight to ten. For development wells, bit runs grew from 113.4 metres a bit in 1970 to 198.2 in 1975. In 1978, the biggest drilling directorate, Nizhnevartovsk No. 1, achieved an average bit run of 262.6 metres, and its best crew achieved 271.5. Bit productivity has grown in spite of the increasing average depth of wells. It has been shown that productivity falls as wells go deeper. In Krasnodar oblast (North Caucasus) for example, a total metrage of 29,950 metres in eight wells drilled to 3,670-3,833 metres required 886 bits (33.8 m/bit) while a metrage of 27,270 in six wells of 4,352-4,794 metres required 948 bits (28.8 m/bit), and 38,315 metres in seven very deep wells of 5,334-5,618 metres required 1,570 bits (24.4 m/bit).

Higher productivity has been accompanied by a significantly faster mechanical drilling speed for development wells. It has grown from 9.21 metres an hour in 1966 to 9.72 in 1970, 11.53 in 1976, and 12.8 in 1978. Thus, mechanical speeds have not only been growing, but accelerating. For exploratory drilling, on the other hand, speeds have declined from 3.78 metres an hour in 1965 to 3.18 in 1970 and 2.96 in 1978, although they appear to have stabilized at this level.

The importance of raising bit productivity stems from the need to reduce the number of round-trip-operations and the time spent on them. More than half of all productive drilling time is taken up by RTOs and the doubling of the average bit run during the 1970s has enabled the share of RTOs in total drilling time to fall from 16.5 per cent to 12.7 per cent for development drilling and from 16.5 per cent to 14.1 per cent for exploratory drilling over 1970-1978. This has enabled the amount of time spent on productive drilling per 1,000 metres to fall by 31.2 per cent for development drilling and 15.8 per cent for exploratory drilling over 1970-1978. It should be noted, however, that improvements in the productivity of down-hole motors has also played a significant part.

Rising bit productivity is due to the constant creation of new, improved designs. During 1971-1975, for example, 35 new types of bit were produced, and in 1980 the Ministry of Chemical Engineering (Minkhimmash), which is responsible for bit production, was elaborating plans for the production of high quality bits so that by 1985 average runs could be increased by 150 per cent to 198.5 metres.

The rolling cutter rock bits (sharoshechnye doloty) are used for 96 per cent of Soviet development drilling and 88 per cent of exploratory drilling. In 1975, 352,000 of these bits were produced compared with 421,000 in 1970. In all, the Soviet Union now produces about 400,000 bits/year.

The best bit plant is said to be that at Drogobych (Western Ukraine) which started up in 1972 to produce the 1-AN range. During the last FYP, more than 30 new models were developed, including 20 models of a new type of bit made from an ultra-hard alloy called "Slavyutich" and developed in conjunction with scientists from the Ukrainian Institute of Ultra-Hard Materials. The bits are designed to drill to depths of 4,000 to 5,000 metres, and are said to operate ten times faster than conventional rock bits. Further research is continuing in a laboratory jointly established with the Lvov Physical-Mechanical Institute. During 1976-1980, output by the Drogobych plant grew by 50.4%, and will expand further over 1981-1985.

The largest bit plant in the USSR is at Kuibyshev. It underwent a major reconstruction in 1978, and in 1982 a new workshop producing 100,000 bits a year with equipment imported from the USA came on-stream.

During the current plan period, new types of core barrel have been widely introduced, together with large-diameter core heads. These have enabled core recoveries of 55-60 per cent to be obtained, i.e. twice as high as before. The KD11-190/80 core barrel used in conjunction with the 6VK-187.3/80SZ and 6VK-212.7/80SZ core heads are said to be of a particularly high quality.

DOWN-HOLE MOTORS

The turbodrill, which is a turbine operated by the pressure of drilling mud on its vanes, was invented in the USSR during the 1930s. However its use only became widespread in the 1950s, when the share of drilling performed by turbodrills, as opposed to rotary drilling, grew from 24 per cent in 1950 to 83 per cent in 1960. This figure declined to 70 per cent in 1969, but has since risen again to 80.9 per cent in 1976.

Production of turbodrills take place almost entirely at Kungur (Perm oblast), and annual output has grown from 978 in 1950 to 6,222 in 1960 and 8,439 in 1965. Reconstruction of the factory led to a dip in production to 6,562 in 1970, but since then it has fluctuated between 9,000 and 10,000. In 1980, 9270 were produced.

As is the case with rigs and bits, the growing productivity of turbodrills has enabled the metrage drilled to rise considerably with a declining output of turbodrills. It can be estimated that the volume of drilling per turbodrill section has risen from 991 metres in 1965 to 1,241 in 1975 and 1,666 in 1980.

These improvements have been obtained partly through the application of higher quality drilling muds. Previous types of mud were highly abrasive, and this led to the vanes of the turbines being worn out rapidly. But there have also been radical improvements in the quality of the turbodrills themselves.

In recent years, production has begun of new types of spindle turbodrills, turbodrills with inclined pressure lines, turbodrills with hydrodynamic braking systems which provide for greater control, and a totally new type of screw-motion down-hole motor. Research work on turbodrills has concentrated on four directions.

a. Improvements through lowering the speed of rotation or through rising the rotation torque.

b. The creation of designs with jet bits and diamond bits.

c. Improvements through the manufacture of castings with greater precision, or from polyamide-12 plastics.

d. The creation of designs capable of suppressing longitudinal and transversal vibration during the course of drilling. These designs involve the independent suspension of the turbodrill in spherical bearings, and are expected to lengthen the life both of turbodrills and drilling bits by providing greater stability. Such a design, the A7ShZ, has been tested in Kuibyshev, and has succeeded in drilling twice the normal metrag for a turbodrill unit.

One of the most important features of a turbodrill is its spindle, and work has been carried out on improving it. VNIIBT has designed a spindle which has increased the inter-repair period by several times when operating with both open and hermetically sealed bearings.

Most modern turbodrills employ centring and stabilising mechanisms which lengthen the life of turbodrills and bits, reduce breakages and improve the quality of drilling. Research is proceeding on the improvement of these mechanisms.

Some Western journalists have claimed that Soviet turbodrills are useless at depths greater than 3,000 metres, and that they drill much more slowly than American equipment.[1] In fact, turbodrills have operated successfully down to depths of 6,000 metres, and comparisons between Soviet turbodrills and modern Western rotary rigs have shown that the turbodrill can drill much faster.

Under a drilling technology exchange scheme between the Ministry of Oil and the Canadian Drilling Research Association, various sizes of modern Soviet turbodrills were tested in portions of six wells in South Alberta. The drilling was carried out in areas where nearby wells had been rotary drilled and it was therefore possible to make comparisons. Generally turbodrills had a faster penetration rate than rotary drills, and in some cases such as the Pre-Cambrian part of the Shell Westcastle well (down to 1,298 metres) the rate was very much faster at 11.1 metres per hour compared with 1.8 for rotary drilling. But

---

1 Time 23/6/80

for the rest of the well, down to 2,879 metres, the difference between the speeds was only 0.9 m/hour in favour of the turbodrill. While the turbodrill proved cheaper to operate than the rotary drill in the Pre-Cambrian part, it was more expensive at greater depths. This was due to a larger number of round-trip-operations because of shorter bit runs.

In addition to the short bit runs, other problems were the inability to survey the well close to the bottom, because the turbodrill was in the way, and the need for more powerful mud pumps on the rigs, with a pump pressure limitation of 155 gk/sq cm proving insufficient. At Samotlor, pressures of 190-210 ats are employed; during 1973-1976, 600,000 metres of wells were drilled with mud pumped at this pressure, and bit runs were increased by 25-30 per cent.

It would seem that the Canadians are interested in the turbodrills advantages under certain conditions, and that they operate much better than some Western journalists assume.

The new types of screw-motion down-hole motor hold great promise for the future. They are more applicable to the modern designs of rock bit with grease-filled floating bearings, they have a high load-transfer capacity, they develop a high rotation torque for slow speeds of 80 to 100 revs/minute, and they are expected to increase bit runs by two to three times. They are being developed by the Turbobur Association, and industrial tests have recently been completed in Tyumen. They speed up drilling by 20 per cent, and it is planned that they should account for 5 mn metres (out of a total of 18 mn) in Tyumen oblast by 1985.

However, the head of Glavtyumengeologiya says that although the prototype has proven "highly successful during tests, the Turbobur plant is too busy fulfilling current orders to find the time or resources to even turn out an experimental batch." He believes that the only answer is to incorporate a turbodrill plant into the new petroleum engineering complex to be set up in Tyumen during the 1981-1985 period.

For the initial stages of drilling, when wells of 345 to 490 mm diameter are being drilled, jet-turbine drills have been developed by VNIIBT. They permit the well to be drilled in one run, and are intended for use in conjunction with serial turbodrills and rock bits. Tests have shown that the jet-turbine drill increases drilling speeds, and reduces costs by 12 to 15 roubles a metre.

OTHER EQUIPMENT

Electric drills

The Russians have been experimenting with electric drills for years, but the most they have ever produced in a year was 220 in 1965. In 1978, production fell to 81, the lowest level for many years, and in 1979 it ceased.

Electric drills are used mainly at the on-shore fields of Western Turkmenia, where great depths, hard rock, anomalously high pressures, well-collapse, blow-outs,etc. make drilling difficult. But drilling with electro-drills has declined from 123,000 metres in 1975 to 82,500 in 1979 because of a failure to acquire an adequate supply of drill motors.

This situation is due to an ill-fated reorganisation of the enterprise producing electrodrills, the Kharkov Electric Motor Plant belonging to the Ministry of the Electrical Equipment Industry. In 1972, it stopped producing electro-drills, and production was transferred to a special experimental production plant which had been set up in Kharkov for the purpose of improving electro-drill designs. This plant was now expected to carry out serial production but proved ill-equipped for this, and most of its experimental work also came to a halt. Even when it was put directly under the jurisdiction of the Kharkov Electric Motor Plant in 1973, the situation did not improve.

Later in 1973, a conference of officials from the oil, gas and electrical equipment industries recommended that the Electrical Equipment Ministry should set up a specialised association for the serial production of electro-drills, but this did not happen until 1978.

In early 1979, the experimental plant ceased production after it had been ordered by the Minister of the Electrical Equipment Industry to produce 570 submersible electric pumps. It is not known how soon the new association will begin producing, but in the meantime there is a shortage of spare parts for electro-drills as well as power cables, etc.

In a letter to Pravda (6/1/81), the foreman of a crew working for the Koturdepinsk drilling directorate of Turkmenneft claimed that the use of electro-drills made for quicker and cheaper drilling, because it saves energy and reduces the usage of drill pipe. Telemetric systems allow the drilling of inclined wells to be carried out, but the production of new electro-drills has been unjustifiably delayed. In reply to the Pravda article, the Ministry of the Electrical Equipment Industry replied that it was working on new models of electro-drills, and that during 1981-1985, twice as many electro-drills would be produced as during 1976-1980.

Drilling muds

The production of drilling muds is organised by the Soyuzneftespetsmaterialy Association, belonging to the Ministry of Chemical Engineering. Its main factory is in the Eastern Ukraine city of Konstantinovsk.

During the Ninth FYP period (1971-1975), oil workers received new types of mechanised oilfield mud-houses for dry muds, 833 complete bases for dry muds, 1,294 complete mud systems for drilling rigs and 130,000 tons of dry barite weighting materials for muds. The Russians say that the quality of chemical reagents was considerably improved, and the demand for bentonite dry muds was fully satisfied.

During the last plan period, the range of muds was extended and their quality improved. Several new types were created. The output of dry barite weighting agents was expanded, and the 75 per cent increase in their prices in 1976 provided the manufacturers with a firm financial base for further expansion. They have concentrated their work on:

    a.   Creating mud systems able to serve a wider range of drilling conditions. New piston pumps like the 15GR with a capacity of 230 HP, manufactured by the Krasnyi Molot Engineering Plant, Groznyi, have proved valuable in this respect.

    b.   Producing better systems for the cleaning and recirculation of muds. This is particularly important in easily pollutable regions like the Caspian Sea and Western Siberia.

    c.   Improving the quality and increasing the output of dry barite weighting agents with moisture contents of less than 3 per cent.

d.   Creating better equipment for the transport, storage and application
     of muds, and mechanising the whole process from the factory to the
     well. The Nalchik Engineering Plant, which produces autocisterns for
     mud transport, has played a major role.

The Ministry of Chemical Engineering has drawn up plans for increasing the output
of muds during the period to 1985. The production of polymerised dry muds,
developed by the Moscow Research Institute of Drilling Technology, has begun,
and new thermostable reagents suitable for drilling holes with bottom temperatures
of up to 240 degrees in high-salt conditions are being developed.

Drill pipe

The USSR has not made much progress in the development of lighter and stronger
drill pipes because the predominance of turbodrilling has rendered this unnecess-
ary. Only with the drilling of very deep wells has the need grown to develop
light drill pipe, and this problem has been overcome by using expensive aluminum
alloy pipe.

At the end of 1980, a total 42 mn metres of oil, gas and other wells had been
drilled with aluminum pipe. The lower column weight has lead to improved safety
for drillers, an increase of 50 to 100 per cent in drilling speeds over conventional drill pipe, and a reduction in power usage of 30 to 50 per cent. The
number of accidents has been sharply reduced, and lighter rigs can be used, which
is particularly advantageous in Western Siberia. It is estimated that savings of
60 mn roubles have been effected by the use of aluminium pipe. It is manufactured
by the "Lenin" Metallurgical Plant, Kuibyshev, and by the Kamensk-Uralskii
Metallurgical Plant in specially designed workshops. However, these plants are
not working at capacity because of reductions in orders by the Ministry of Oil,
the main user of aluminium pipe, in spite of a growing demand by drilling enterprises. This is because the Ministry of Oil is afraid to reduce its orders for
steel drilling pipe from other plants, in case it cannot increase them again
in the future.

In 1966, production began of a new design of high-tensile pipe, with stabilising
bands, of types TBVK and TBNK intended for low-speed drilling. TBVK means "pipes
for drilling upper strings", and TBNK is for lower strings. It is claimed that
drill string breakages have practically ceased, although during the three years
1976-1978, Tyumen oil workers suffered 65 accidents and 157 breakages of drill
pipe, leading to a failure to produce 2 mn tons of oil. The head of Glavtyumengeologiya has complained that drill pipe is of poor quality and breaks frequently.
In a pre-Congress letter to Pravda, he urged that the guidelines for the Eleventh
FYP should include the need to accelerate the production of reliable special
types of steel pipes with improved threads and anti-corrosive coatings, and that
the output of aluminium pipe should be increased.

During the last FYP period, the USSR carried out:

1.   the production of TBVK and TBNK pipe of diameter 89 to 140 mm with a
     strength of 95 kg/cm;

2.   the preparation for serial production of pipe with welded ends with
     fluidity limits of up to 75 kg/cm, and light alloy pipe with conical
     stabilising bands of types LBTVK-129 and LBTVK-140 (light alloy drill
     pipe for upper strings of diameter 129 and 140 mm);

3.   the production of a new type of drill collar, the UBTS - 146-229,
     with diameters of 146 to 229 mm and leading pipes with conical
     stabilising bands of type TVKP, with diameters of 112, 140 and 155 mm;

4.   the creation of new types of joints with threads of increased durability of type MK.

5.   the manufacture of heat-treated pipe designed for super-deep wells. It is said to enable deep drilling to be carried out 20 per cent faster than before, with the average life of the pipe extended from 400 to 2,500 hours.

Casing pipe

The USSR produces about 1 mn tons of casing pipe a year. As with drill pipe, there have been persistent problems with inadequate supply and poor quality. Soviet casing pipe tends to be weaker and heavier than Western pipe, weighing about 50 kg a metre on average. Between 4 and 6 per cent of all casing pipe is rejected because of poor quality threads, a lack of internal and external facing and metall- urgical defects. Siberian oil men have even set up a special plant in Nizhnevar- tovsk to repair defective pipes.

During the current plan period, the Russians have started to remedy these problems by producing new designs of collared joint casing pipe with trapetsidal threads of type OTTM, flush collarless pipes of type OG-1 with high tensile joints of type OGTG-1, and also collarless pipe with high tensile joints of type TBO.

Casing pipe comes in ten basic sizes with diameters ranging from 127 to 426 mm. Wall thicknesses vary between 7 and 11 mm for pipes of 127 to 245 mm. The most important pipe-producing plant is the Almetyevsk spiral-seam pipe plant in the Tatar republic. It produces 330,000 tons of pipe a year by high-frequency welding at a rate of 20 metres a minute compared with 1.2 metres a minute under traditional electric arc welding methods. Its spiral seams mean that considerably less steel is required for the pipe than when direct seams are used. It has produced 2 mn tons of spirally welded pipe during the last 15 years; the first million tons took 12 years to produce, and the second million required only three years.

The plant is now being extended, and two new mills will raise its capacity to 700,000 tons/year. The third mill will specialise in pre-insulated pipe covered both internally and externally with anti-corrosive substances; this should extend its period of service by several times. Initially, 50,000 tons of this pipe will be produced a year, and new developments have been assisted by the 68 per cent increase in the price of drill and casing pipe.

Other important producers of drill and casing pipe are the Severskii Pipe Works, south of Sverdlovsk, the Chelyabinsk piperolling plant, the Libknecht Nizhned- neprovsk piperolling plant (Ukraine), which produces 66,400 tons of mostly 245 mm diameter pipe a year, and the Zhdanov Pipeworks, also in the Ukraine.

Cementing and well repairs

Most associations include a directorate for repairing wells, and in 1980 the USSR had 1,030 crews engaged in current repairs and 940 in capital repairs. They carry out 250,000 current repairs and 940 in capital repairs. They carry out 250,000 current repairs and 20,000 capital repairs annually. A large share of these (120,000 current and 3,100 capital repairs) are carried out at the aging on-shore fields of Azerbaidzhan.

The repair directorate belonging to Nizhnevartovskneftegaz Association in Western Siberia was expected to repair 522 wells in 1979, but in spite of this, 2.9 per cent of all Western Siberian wells were idle during 1979 compared with 2.6 per cent in the USSR as a whole, and only 1.1 per cent in the Bashkir republic, where well repair crews are highly trained and well organised.

There has been a significant improvement in the quality of repair jobs in the
last few years, with inter-repair periods for most types of well growing significantly.

Table 26

Inter-repair Periods for Oil Wells, days

	1976	1979	Increase, per centa year
Free-flowing wells	773	1,014	9.5
Pumped wells:			
sucker rod pumps	230	240	1.5
electric submersibles	91.3	98.3	2.5
Gaslift wells	516	472	-3.0

Source: Mingareev et al p. 106

The quality of repairs needs to be raised, and 7 per cent of all repairs in 1979
were for the second time. Well repairs are generally necessitated by poor
cement jobs, and special attention has been paid to the question in the Tatar
republic where cement jobs have been improved by the following:

a.   A series of measures to ensure that the cement is squeezed to the top
     of the annulus (i.e. the well mouth) including the use of improved
     types of cement and the reduction of pressure in the annulus. These
     were applied to 2,433 wells.

b.   The use of improved packer filters designed by the local research insti-
     tute. These were used in 206 wells.

c.   The use of buffer fluids, for the flushing of the mud cake, and oil-
     based solvents.

d.   Silicate processing of the productive interval of the section before
     cementing. These last two methods were employed at 2,000 wells.

In West Siberia, problems arise due to the low temperatures which have a negative
influence on cementing. The Iskitim cement plant is the principal supplier for
Siberian fields, producing more than 100,000 tons of special cement a year. But
its quality must be improved, because wells fall into disrepair faster in
Western Siberia than anywhere else.

During the last five years, widespread use has been made of geophysical methods
of testing cement jobs, i.e. with radioactive and acoustic cementometers. The
VNIIneftepromgeofizika Institute has designed and begun the serial production
of the SGDT-2 device for revealing defects in cementing. THE VNIIKRneft
Institute, jointly with the Yuzhmorgeo (Southern Marine Geophysical) Association
has designed a phase correlation logging unit BFK which can be used in conjunction
with its acoustic cementometer AKTs to obtain data on the extent of contact between
cement and casing on the one hand, and cement and the well-wall on the other, and
also to reveal the intervals over which uncemented casing is in contact with the
well wall. At the moment, these devices are used mainly in those regions which,
for a variety of reasons, have bad records of cement work - Tyumen and Tomsk in
Western Siberia, Stavropol and Krasnodar in the North Caucasus, Orenburg in the
Urals and Mangyshlak in Kazakhstan.

Well repair equipment is produced by several plants. The Sardarov plant in Baku has produced the USSR's most powerful machine for well repairs. It is an entire repair enterprise mounted on the chassis of a heavy duty lorry. The Rustavi crane produces 750 cranes for oilfields each year and hoist equipment is produced by the Leitenant Shmidt plant in Baku, whose latest model has a lifting capacity 50 per cent greater than previous designs. Instruments for winding and trans- porting cable are made by the Ishimbai Engineering Plant, and the Geomash plant, Shchigri, makes a unit which installs masts for well repairs. The Azinmash plant, Baku, makes circulating swivels for flushing sand plugs, and the Nefteburmashremount plant, Kalush (Ukraine), makes special packers.

However, the USSR still needs to import equipment. Rumanian repair units have recently been imported for use in Turkmenia, where they will repair 4,000 metre wells.

Well cleaning

In some regions, wells tend to become blocked by resins, asphalts and paraffin deposition. Several types of cleaning processes are used, including thermal processes, surfactants, sulphuric acid, hydrochloric acid, thermo-acid processes etc. In 1978, 15,000 well bottoms were cleaned, providing an additional 2.3 mm tons of oil. After thermal processing (2,300 operations) the increase in the extraction of oil is greater than for other methods, although thermal pro- cessing is not yet employed in many smaller regions.

New solvents for flushing wells after paraffin deposition are being continuously developed by the chemical industry.

Drill stem testing equipment

The number of drill stem tests in the USSR has grown from 600 in 1967 to 3,900 in 1977, and the share of tests successfully accomplished has risen from 78 per cent to 95 per cent. New well-logging instruments introduced in the Turkmen republic are designed to calculate strata porosity at depths of 6,000 metres under pressures of 1,000 ats. They are supplied by the Kiev Instrument Works. Other suppliers include Azinmash of Baku, which produces the A-50 instrument, and the Kasimov Engineering Plant of Baku which has produced the prototype of a new model, the LS-6 designed to work at 7,000 metres. The Kasimov plant has also designed a complex mounted on a tracked vehicle for testing wells to 4,000 metres in Western Siberia. The Volodarskii Engineering Plant of Baku makes special packers for drill stem testing, and the "Nefteavtomatika" plant of Bulgulma makes equipment suited for the Urals-Volga fields.

Oil preparation installations

These are designed to remove water, salt and other impurities from the oil prior to its transportation, and almost all new installations are being erected in Western Siberia. Soviet produced plants normally have capacities of 6 mn tons/ year, such as those at Samotlor.

In September 1977, the purchase of 26 thermo-chemical installations from East Germany was announced. They each have a capacity of 3.5 mn tons/year, giving total capacity of 91 mn tons/year. The units were designed by the USSR, but manufactured in East Germany and by late 1979 they were being installed at the Surgut oilfield.

Oil processing in Western Siberia is carried out by the Zapsibnefteavtomatika Association, which is part of Glavtyumenneftegaz. At the Nizhnevartovsk oil collection point, where oil is prepared for transmission through the trunk pipeline system, there are 16 preparation complexes.

A decision has been taken to re-equip the treatment plants in the Tatar region. A new model designed by the "Neftekhimpromavtomatika" Special Design Bureau, and automatically controlled by a micro-computer, has been installed at Almetyevsk.

The complexes are produced by Uralmash of Sverdlovsk and several other engineering plants in the Urals and Azerbaidzhan. Anti-flare devices are manufactured by the Nevskii Engineering Plant in Leningrad.

Gaslift equipment

The 1980s are likely to see far greater use made of the gaslift method, particularly in Western Siberia and the Eastern Caspian Coast. This is because gaslift has proved more suitable for the extraction of heavy oils, compared with waterflooding, and because at some deposits with high salinity characteristics waterflooding is proving impracticable due to the short inter-repair period for submersible pumps - only 98 days in 1979 compared with 472 for gaslift wells.

At Samotlor and Fedorovsk in Western Siberia, for example, the average inter-repair period for pumps blocked by salt accumulation is only 60 days, and the average for Western Siberia is 90 to 120 days, although at deposits like Ust Balyk and Zapadno-Surgut it amounts to 300 days. The urgency behind the transfer to gaslift at Samotlor is explained not only by the salinity, but also by the low capacity and poor efficiency of Soviet pumps. It is expected that 200 wells at Samotlor will yield 2,000 to 2,300 cu m of fluid (oil & water) a day, with a further 1,000 yielding 1,000 to 2,000 cu m/day. Soviet pumps have capacities of only up to 700 cu m/day, and a 1,000 cu m/day pump has only just been tested. The rate of usage of pumped wells is only 70 to 90 per cent, and the Russians recognise the urgent need to produce pumps able to work for up to three years between repairs, providing a usage rate of 90 to 95 per cent.

Gaslift has several other advantages. It permits use to be made of casing-head gas until gas collectors and processing facilities can be built, it makes for higher oil yields per well, and it reduces the number of servicing personnel required.

Gaslift has been used in Western Siberia for ten years. In mid 1969, a contract was signed with Camco of Canada for the installation of a gaslift system at Pravdinsk deposit. The first parts arrived in 1970, and during 1971-1972 the industrial testing of the underground equipment (tubing, valves, etc) was carried out using air from a mobile compressor instead of casinghead gas. During the next four years, wells were converted to gaslift at the rate of 50 a year, so that by the end of 1975 200 gaslift wells were operating. The number has continued to grow, although much more slowly (in 1978, Pravdinsk had 222 gaslift and 110 free-flow wells), and complaints have been voiced recently that deliveries of equipment have fallen short of requirements.

In 1974, a system was installed at Zapadno-Surgut which was said to have the same advantages as gaslift over the traditional methods of the mechanical extraction of oil by centrifugal submersible pumps and sucker rod pumps. A hydro-piston pumping unit pumped fluid extracted from the deposit back down the well, and this lifted heavier crude to the surface. By 1976, it was undergoing industrial tests, and was said to work well for clusters of inclined wells.

At Samotlor, gaslift was introduced in 1975, when nine wells were tested with
the gas being forced back under its own pressure, thereby obviating the need
for compressor units.

However, the volume of Tyumen oil raised by gaslift methods remains small, and
in 1978 only 4 per cent of the region's output (i.e. 9.8 mn tons) was produced
in this way from 6 per cent of its wells, although it is said that these figures
will rise considerably in the future. One authoritative source says that
gaslift will be employed at more than 40 deposits, including Fedorovsk, Bystrinsk,
Tagrinsk, Lyantorsk and Varyegansk as well as Samotlor, Pravdinsk and Fedorovsk.

The reason for the slow rate of introduction of gaslift, in spite of its advantages, has been the high capital cost (particularly of the installation of
compressor units) and the low level of efficiency. The increased control prices
for crude oil projected for 1981-1982 should encourage the greater use of gaslift
if the low efficiency problems can be overcome.

Consequently, research workers have been looking into ways of raising the level
of efficiency and their work has proceeded in three directions.

   a.   By reducing pressure losses through gas seepage. Experimental work has
        been carried out since 1976 at Pravdinsk with a gas dispersal install-
        ation which splits up the gas phase so as to enhance its lifting
        capacity and reduce the requirements of gas per well.

   b.   By increasing the depth to which gas is injected. At Pravdinsk, a
        new method of arrangement and operation of valves in inclined wells
        has been successfully tested.

   c.   By improving the operation of gaslift wells by optimising the distri-
        bution of gas between them. An experimental system has been employed
        for 20 wells at Pravdinsk.

So as to reduce the capital cost of gaslift, two schemes of "autonomous non-
compressor gaslift" have been under test at Samotlor since 1975. They each
operate on an artificial island from which a cluster of inclined oil-wells has
been drilled. The new scheme obviates the need to drill two special gas wells,
erect a facility for the preparation of gas, lay high-pressure gas pipelines,
and build special gas distribution batteries. The employment of this method,
compared with traditional gaslift, leads to savings of 45,400 roubles a year,
and an additional 540,400 tons of oil were obtained from eight Samotlor wells
over four years.

Of the three measures designed to raise efficiency, the first is the most
effective, providing savings of 18,600 roubles a year. Between them, they enabled
an additional 706,600 tons of oil to be obtained from the 278 wells at which they
were employed over the five years 1974-1978. Of the 42.2 mn tons of oil obtained
by gaslift in Siberia over this period, the new measures for raising efficiency
and cutting capital costs accounted for 1.25 mn tons.

A problem which arises specifically where gaslift is used in conjunction with
waterflooding is that of hydrate formations accumulating in the gaslift tubing.
Sometimes, hydrate plugs up to 330 metres long are formed, completely blocking
the tubing. In 1976, 28 wells were blocked in this way at Pravdinsk, and
special scraper rods were employed for the removal of unconsolidated plugs.
Where the hydrate has become consolidated, it has to be removed by steam or hot
water injected by high-pressure jets.

In 1978, it was decided to convert Samotlor and Fedorovsk to gaslift. The equipment was initially ordered from the US firm Brown and Root for $192 mn, but this firm could not fulfill the contract. Consequently it was awarded to Technip and Creusot Loire of France for $201 mn.

The first phase of Samotlor's gaslift plan covers the three years 1981-1983. It envisages the conversion of 1,500 wells and the installation of 14 well-head compressor stations, 1,000 kms of pipelines for carrying the gas and 250 distribution centres operated with the aid of computers. With the completion of the second phase in 1985, a planned 40 to 50 mn tons a year will be obtained by gaslift, thereby releasing 500 workers for other jobs. Samotlor and Fedorovsk will have 4,025 gaslift wells between them.[1]

Some of the equipment will also be made in the Soviet Union, by the Kazan Compressor Plant in conjunction with the Bugulma Production Association of Oil Refining Engineering. The compressors being built for Samotlor are of 6 MW capacity and pump gas at 110 ats. The Kazan plant will manufacture 34 compressors during 1981-1985.

During the 1970s , the volume of gas and air pumped into wells rose from 340 mn cu m in 1970 to 501 mn cu m in 1979, due mainly to its greater use in the Kazakh, Turkmen and Azerbaidzhan deposits. In 1980, there was a significant fall to only 421 mn cu m, due primarily to the closure of many wells in the North Caucasus oilfield where gaslift has been used extensively in the past.

Pumping equipment

Despite the plans for a considerable expansion of gaslift during the 1980s , the use of waterflooding in conjuncti on with mechanical pumping will remain the most important method of oil production until well beyond 1990. During the four years 1976-1979, more than 4,000 wells were converted to mechanical pumping, and Western Siberia alone will have up to 13,000 pumped wells in the near future.

The standard range of pumping units is built by the Azinmash plant of Baku, and consists of nine basic models and 11 modifications, with payloads of 1 to 20 tons and torques of 100 to 12,000 kgs per metre. The most important models are the SKN2-615, SKN3-1515, SKN5-3012 and SKN10-3315, where the first number represents the payload, and the second the torque. These are employed at 60 per cent of all wells with sucker rod pumps.

New non-balanced models with payloads of 3.6 to 12 tons have been devised; these are said to be smaller and lighter than balanced models of similar capacity. New designs with pneumatic counter-balancing and hydraulic drive are being tested and all new models will have two-stage cylindrical gears.

Both sucker-rod pumps and electric centrifugal submersible pumps are used, with the former being used specifically in low yield or high salinity deposits. But electric pumps have a far greater capacity, and producing more and better electric pumps will be a key task for Soviet petroleum engineering during the 1980s.

In 1975, the Soviet oil industry had 92,700 wells, of which 68,000 produced oil, and the rest were injection and observation wells, or exhausted oil wells. Some 54,000 wells were pumped with 9,100 being worked by electric pumps and the rest by sucker-rod pumps. Between 1976 and 1980, the number of wells worked

---

1 Economist Intelligence Unit, Quarterly Energy Review, USSR & Eastern Europe. 1981 No. 4

with sucker-rod pumps grew from 43,372 to over 45,000, and their average yield in 1980 was 4.5 tons a day, with mechanically pumped wells giving 72 to 74 mn tons a year. In 1980, the USSR had a stock of about 60,000 sucker-rod pumps of which 75 per cent were operating.

In Western Siberia, they had 701 electric pumps and 460 sucker-rod pumps operating in 1975 (176 and 98 respectively in 1971), producing 16 mm tons of oil a year. The importance of electric pumps stemmed from the very high fluid-lifting requirements of Siberian wells, and it proved necessary to employ them in spite of the salt deposition problem. Research has been carried out into the development of new pumps able to withstand salt deposition for longer periods. In 1972, a new type lasted twice as long as usual at the Shaim field, and in the same year a new design at Ust Balyk lasted for 18 months before repairs were necessary. The latter was suspended on a cable instead of using tubing, and proved very good for inclined wells. It could be raised for repairs much more easily.

The fluid-lifting requirements at Samotlor are so great that Soviet-made pumps are unsuitable. Reda pumps of capacity 800 to 1,000 cu m/day have been imported from the USA, and according to the CIA 1,210 of these pumps were to have been imported by the end of 1977. However, supplies were limited to 30 a month, and the USSR has been working on producing its own high capacity pumps. It is claimed electric submersible pumps of 1,000 to 1,400 cu m/day have been designed, and that the Almetyevsk electric pump enterprise has begun serial production.

In the Ural-Volga region, fluid-lifting requirements per well are not so great. In 1977, 172.2 mn tons of fluid (132 mn tons of water and 40 mn tons of oil) were pumped from the Bashkir republic's 12,199 wells, of which 3,024 were equipped by electric pumps and 9,045 by sucker-rod pumps. An average of 43.6 cu m/day were pumped, including 146.6 from wells with electric pumps and 13.5 cu m/day from wells with sucker-rod pumps.

Between 1971 and 1978, the number of electric pumps grew from 1,365 to 3,024 at a rate of 12.0 per cent a year, while the number of sucker-rod pumps grew from 6,606 to 9,045 at 4.6 per cent a year. The number of sucker-rod pumps is growing because they are adequate for low yield wells, and because the faster introduction of electric pumps is hindered by the presence of hydrogen sulphide and gypsum. Pumps working in wells with hydrogen sulphide generally need repairs after 200 days, and in wells with gypsum deposits, the pump is irreparable after working for this period.

The production of better quality pumps is helping to overcome this problem. In 1971, wells with electric pumps worked, on average, for 188 days between repairs, and in 1979 for 273 days. The improvement in sucker-rod pumps was even more striking, from 137 days to 257 days between repairs; this has been helped by better packing which has lead to fewer rods being deformed during transport and storage, and then breaking on use. The number of well-days per rod breakage has more than doubled from 429 to 995 in only six years.

Another reason for the better performance of rods has been the increasing use of new designs of narrow pump of diameter 28 to 38 mm which are less likely to break the rods than the traditional types of 56 to 68 mm. Their number trebled from 1,100 to 3,300 over 1971-1977 while the number of traditional pumps remained at 5,500.

A further factor improving sucker-rod pump performance has been the painting of rods and tubing with epoxy resin or enamels. This effectively reduces the accumulation of paraffin resin, and thereby reduces the load on the equipment. Paraffin deposition occurs in 2,500 Bashkir wells (20 per cent of the total) and salt in 500 wells. Since 1974, various solvents, produced both domestically and by foreign firms like Petrolite, have been used as well as heat treatment to overcome the problem.

Since 1977, a number of measures have been taken to reduce rod breakages even further. The work of rods has been studied as a function of the type of steel, the plant of origin, the diameter of the rod, and the environment in which it works (salinity, temperature, extent of waterflooding, etc). All wells with breakages averaging three or more a year (about 300 wells) are equipped with special high-tensile steel rods. A control centre for checking rods before they are used and a central repair depot have been set up at Neftekamsk, and repair bases for tubing have been created at Oktyabrskii and Neftekamsk.

At electric pump wells, only 14 per cent of breakdowns are due to pump failure, with a further 17 per cent due to defective cables. This may be because electric pumps at Bashkir deposits are low capacity; it is with the large capacity pumps necessary for Western Siberia that the Soviets are having problems.

The Bashkir republic is an example of where operating conditions are easy and where the demands on pumps are not too great. The neighbouring Tatar republic and Kuibyshev oblast also claim good records in pump operation - 327 days between repairs for electric pumps and 317 days for sucker rod pumps. In Western Siberia and the Komi republic, on the other hand, the inter-repair period is only 95 to 143 days for electric pumps, and in the Komi and Chechen-Ingush republics, the Emba field of Kazakhstan and the Turkmen republic, sucker-rod pumps last for less than 85 days on average.

One factor, so far largely ignored in the USSR, which can contribute to improvements in the effectiveness of pumps, is the correct choice of pump for the conditions. In Azerbaidzhan, 56 of the 280 electric submersible pumps are of the ETsNI-6 type, with inter-repair periods ranging from 40 to 73 days. The rest are in the ETsNI-5 range, and their inter-repair period can average 122 days. The replacement of the former by the latter has been recommended.

Sucker-rod pumps are produced mostly in Baku by the Leitenant Shmidt and Azinmash plants, by the Lenin Engineering Plant (Perm), the Ochersk Engineering Plant (Perm oblast), the Lebedyan Engineering Plant (Lipetsk) and the Bobruisk Mechanical Engineering Plant of Belorussia. They come in type sizes with diameters of 28 to 93 mm and a plunger a range of 600 to 6,000 mm, and have capacities of up to 300 cu m/day.

Electric pumps are made by the Dzerzhinskii Engineering Plant in Baku and the Kharkov Electric Motor Plant, and the most modern pumps of 1,400 cu m/day capacity are made at a new plant in Almetyevsk (Tatar republic) which began operating in 1980. Tubing for pumps is made at the Sumgait Engineering Plant in Azerbaidzhan.

The Yuzhgipromash plant of Berdyansk makes pumps designed to pump water down injection wells. It makes two types, the "2000" and "3000" with capacities of 2,000 and 3,000 cu m/day of water. The "2000" can deliver water to a depth of 1,000 metres and the "3000" to 1,400 metres. The plant has recently modernised the pumps, adapting them to use reservoir water as well as pure water, and is preparing to extend its range with the production of 4,000 cu m/day pumps.

Oil gathering networks

During the last decade, the Ministry of Oil has been carrying out the reconstruction of the oil gathering network at old deposits, with the aim of reducing
losses of oil during collection and storage prior to transportation. During
the period 1971 to 1975, losses were reduced by $2\frac{1}{2}$ times, and amounted to 1.35
per cent by 1975. During the current plan period, this improvement continued
to the extent that only 0.68 per cent was lost in 1978, and the 1980 plan
target of 0.65 per cent was surpassed, with a loss rate of 0.64 per cent.

This improvement has been brought about by the use of hermetically sealed
single-pipe systems with capacities of 1, 3, 6 and 9 mm tons/year. In Western
Siberia, all deposits are equipped with such systems, and even at long established
fields like the Bashkir republic 94 per cent of all wells are connected by them.
It is planned to reduce the volume of oil extracted by the "open system" to
less than 8 per cent, with particular attention being paid to regions where
losses still exceed the norm: these include Azerbaidzhan and the Tatar and
Bashkir republics.

Recent developments have included the design by the Ministry of Chemical
Engineering and the Machine Tool Ministry of modular units for the collection,
oilfield transport and preparation of oil, gas and water, including fully sealed
storage tanks. Examples introduced recently include the complex with a 20,000
cu m storage reservoir at Lokosovsk, and those with 10,000 cu m reservoirs
at Agansk and Pokacheva deposits in Western Siberia.

Important contributors to the new equipment are the Saratov Petroleum Engineering
Plant which annually manufactures 400 special valves intended for regulating
pressures in oil storage tanks, and pumps for the pre-transport preparation of
oil, and the Baku plants belonging to Soyuzneftemash, which make well-head
equipment.

Gasfield equipment

The most important item of gasfield equipment is the gas processing installation
which removes condensate, hydrates and other impurities from the gas prior to its
transportation. It consists of a separator (to clean the gas), an absorber (to
dry it), and a filter, and is particularly necessary at gas condensate deposits
such as Urengoi in Western Siberia.

The Tenth FYP called for the introduction of modular automatic units with
capacities of 3 and 5 mn cu m/day which would include cooling systems for the
handling of very hot gas. These units are now being produced on a serial basis,
with coolers made by the Lauriston Engineering Works of Tallin. The new deposits
in Western Siberia, however, have such large well yields (only 70 to 80 wells are
needed for the production of 100 bn cu m of gas at Urengoi compared with 1,500
wells in the Ukraine) that processing installations of far greater capacity can
be employed without excessive expenditure on gathering networks. At Medvezhe,
the first four of the nine plants had capacities of 13.5 mn cu m/day. The
first four plants at Urengoi each process 27 mn cu m/day and the fifth plant,
installed in late 1980, has a capacity of 40 mn cu m/day. The new Yamburg field
will be served by seven such plants with a total throughput of 280 mn cu m/day.

These large plants are being introduced at rapid rates, with plans calling for
ten plants of total capacity 270 mn cu m/day to be built in Siberia during the
three years 1978-1980. While there have been long delays in the past, the
growth of modular construction methods is enabling work rates to be raised.
Originally, the plants had 16 metre columns, and weighed 80 tons. They were

delivered in two parts, but since 1978 a new model with the same daily capacity and performing the same function, but with 8-metre columns and weighing only 20 tons, has been manufactured.

Modular construction enabled Urengoi's first processing plant to be installed in four months compared with the projected two years. It is likely that in future these plants will, like Urengoi-4, be combined with cooling equipment to permit the transport of cooled gas.

One of the main factors delaying the installation of processing plants is the failure to build artificial islands (of the type discussed in the section on rigs) in good time, and this in turn stems from the shortage of roads. At Urengoi, 12 mn cu m of sand should have been collected from the beds of lakes and rivers for island construction in 1980, but only 7 mn were acquired due to the shortage of dredgers. For some gasfields, the sand must be brought hundreds of kilometres because the local sand lies at depths of up to 20 metres. At Urengoi, "sandwells" have been drilled, and a sand/water mixture is pumped up with the use of airlift.

The use of large capacity processing plants often necessitates a more extensive gathering network. An important target of the Tenth FYP was the creation of high tensile (up to 60 kg/cm^2) gathering pipes of diameter 200, 250 and even 350 mm. They were to be insulated internally and externally at the factory, and form part of a hermetically sealed system together with new types of well-head installation. New types of insulation have been an important feature of both oil and gas industry plans in recent years.

At some gas deposits, treatment plants able to scrub the gas of sulphur are required, and the two largest in the Soviet Union are at Orenburg (Urals region) which has three plants of 15 bn cu m/year each, and Mubarek in the Uzbek republic with a total capacity so far of 15 bn cu m/yr. While the Orenburg installation was imported, the Mubarek plant has been built from domestically produced equipment.

Modular construction

The installation of large units such as pipeline pumping and compressor stations, waterpumping stations, and oil and gas preparation units has proved costly and time-consuming, particularly in the difficult conditions of Western Siberia. The practice of completing them at their factory of origin, and transporting the entire completed unit, or a small number of large sub-units, to Siberia, is now widely employed.

The first pre-assembled pumping station was installed on the Shaim-Tyumen pipeline in 1970 and only six months were required before it came on-stream compared with the usual two years. However, it was five years before Sibkomplektmontazh, a specialist organisation for building and supervising the installation of modular units, was set up by Minneftegazstroi.

In 1978, the first modular cluster pumping unit consisting of one complete element was installed at the new Siberian deposit of Pokacheva. The unit was built in the experimental plant of Sibkomplektmontazh at Tyumen. Weighing 250 tons, and including all the necessary equipment, it compared with the eight to twelve separate elements that previously made up a pumping unit, and which had to be asembled on-site. With the new unit, the only on-site work necessary is the digging of the foundations and the laying of power and communications lines. Serial production of the new unit has begun and seven were built last year, including some models weighing 400 tons.

During the last FYP (1976-1980), Sibkomplektmontazh installed nine gas preparation plants, 12 oil preparation plants, 49 oil-pumping stations, 26 gas compressor stations, 30 cluster pumping stations for oilfields and 652 boilers. The Tenth FYP target of 459 mn roubles work was fulfilled 11 days ahead of schedule. During the Eleventh FYP, the range of items is to be extended and the association is working with the research organisation Giprotyumenneftegaz in this direction.

Today, units of 500 tons are being manufactured, but the problem of how to transport them becomes more acute. They are sent into Siberia by river in summer, and then have to wait for winter, when they can be hauled by transporters to their destination. The most recent type of transporter is that designed by Bryansk Diesel Works, consisting of a special traction unit and a large trailer with an unloading mechanism. This process defeats the original purpose of modular construction, and, until all-weather roads have been built, hovercrafts seem to be the only answer. A hovercraft able to carry a 20-ton load was successfully tested in 1979, and began work at Urengoi in 1980, and Sibkomplektmontazh has produced its own hovercraft with a carrying capacity of 400 tons, although it has to be towed by 3 tractors. It is now being tested on the Nizhnevartovsk oilfield.

Under traditional methods, 1,000 workers were needed to build one pumping station in 18 to 20 months. With modular construction, 45 to 50 workers can build four to five pumping stations a year. It is expected that units of more than 1,000 tons will eventually be built, but an increasingly common complaint is that modular construction of auxiliary equipment is being ignored. With cluster well drilling, hermetically sealed oil collection systems, compressorless transport of casinghead gas, and the transport of oil saturated with gas, there are endless possibilities for the modular construction of items which are very small, but would nevertheless enable a great amount of assembly time to be saved.

It is further argued that savings of money and time from the modular construction of large units are being eroded, simply because such large units are being built. The questions of organisation and management have not been studied sufficiently.

CHAPTER 7

Production Problems

ORGANISATION

The guidelines of the Tenth FYP called for the efficiency of drilling work to be raised and the average period of well construction to be reduced by 25 to 30 per cent, i.e. from 126 days in 1975 to 92 in 1980. By 1978, it took 103 days to complete the average well, and by 1979 it was claimed that the 1980 target had been fulfilled a year early. These figures refer to exploration and development wells drilled for oil and gas by the ministries of Oil, Gas and Geology. For the Ministry of Oil alone, the average completion time has been reduced from 116 to 83 days, and for development wells alone from 74.3 to 52 days. There has been very little change in the average completion time for exploratory wells drilled by the Ministry of Oil, from 328 days in 1975 to 322 in 1980.

While much of the reduction in well completion time can be attributed to faster drilling, the creation of easily erected rigs, and the introduction of new types of bits, downhole motors, muds, and drilling and casing pipe, there is no doubt that better organisation has also played a part.

Better organisation has contributed to a slowdown in the rise in drilling costs. The shift in the centre of gravity of the oil industry to Western Siberia led to a dramatic rise in the cost of exploratory drilling from 112 roubles a metre in 1960 to 149 in 1965 and 252 in 1970. It rose further to 303 in 1974, but has since started to decline to 279 in 1978. A similar rise was expected for development drilling costs, but the biggest increase in costs came before Siberian drilling assumed its current importance, i.e. during the 1960s, when costs per metre rose from 49 roubles in 1960 to 85 in 1970. During the following eight years, costs rose by only 28 roubles to 113 roubles a metre, in 1978, and may be declining now. But as more remote fields with complex geologies are drilled, costs may begin to rise again, and it is significant that while drilling at Samotlor costs only 66 to 79 roubles a metre, at the more difficult Varyegan group of deposits it amounts to 161 to 170 compared with 90.7 for Western Siberia as a whole in 1975.

In 1978, less than a third of the time taken to construct a well consisted of drilling time, which included the actual drilling process (18.5 per cent of the time) and round-trip-operations (12.7 per cent). Nearly as much time was taken up by auxiliary work (electrical logging, orientation of drilling tools, the preparation and pumping of drilling muds etc.), and nearly 40 per cent of well time consisted of non-productive work, the overcoming of complications, accidents and breakages and idling. The picture was similar for exploratory wells, with productive work amounting to 34.7 per cent (drilling 19.2 per cent, round-trip-operations 14.1 per cent), auxiliary work 26.4 per cent, non-productive work 13.6 per cent, overcoming complications 7.9 per cent, accidents and breakdowns 6.4 per cent and idling 11 per cent. For both development and exploratory wells, there has been little change in these figures since 1965 although the average well construction time has been considerably reduced, and this suggests that technological and organisational improvements have had a more or less equal impact.

Organisational improvements have been most noticeable during the last few years. The ratio of uncompleted work to capital investment in well construction reached a peak in 1976, and at the beginning of 1976 only 39.8 per cent of development wells under construction were being drilled. Another 10.5 per cent were awaiting drilling, 15.1 per cent were held in reserve, 2.6 per cent were awaiting closure, 24.8 per cent were being tested and 7.2 per cent were idle for other reasons. For exploratory wells drilled by the Ministry of Oil, the situation was even worse, with only 26.7 per cent being drilled, 53.5 per cent awaiting transfer to production organisations, 2.3 per cent held in reserve, 3.8 per cent awaiting closure and 13.7 per cent idle for other reasons.

During the next two years, a radical change took place in accordance with the plan to reduce well-construction time by 25 to 30 per cent. The share of wells awaiting drilling was reduced to 7.1 per cent, and the share of wells under test was nearly halved to 13.2 per cent. The period during which exploratory wells awaited transfer to production organisations was greatly reduced, and in some regions completely abolished. Thus the share of wells awaiting transfers was reduced from 53.5 to 4.1 per cent, while the share of wells being drilled doubled to 53.2 per cent.

These improvements enabled the number of development wells being drilled to increase by 19.8 per cent while the rig inventory declined. The average period of development well construction was reduced by 25.6 per cent during the period 1976-1979, thereby meeting the FYP target, and for exploratory wells it was reduced by 13.5 per cent. The volume of uncompleted work fell back to an acceptable level.

The reduction in well construction time would have been greater had it not been for an increase in the time spent on auxiliary work in many regions. This was caused by deeper holes, more complicated geology, and a greater share of inclined wells. But plans to reduce auxiliary work time are under way. These include the introduction of more efficient electrical logging processes, the use of modern methods of preparation and cleaning of muds, a better coordination of mud preparation with drilling, the optimisation of the flushing and reaming processes, and a faster installation of blow-out preventers.

Considerable scope remains for further organisational improvements. This can be ascertained by comparing the work of an association's best drilling directorate with that of the association as a whole. Thus the average crew employed by the Yuganskneftegaz Association (Western Siberia) drilled 33,743 metres in 1979, while the crews working for its best drilling directorate, Nefteyugansk No. 2, averaged 52,397 metres. This situation was typical of all associations. There are three main reasons for this wide variation between the best and the average drilling performances.

First, the shortage of specialised engineering and technical service organisations means that drilling crews often have to perform specialised work themselves. Specialised services make for better well design, the choice of more suitable rigs with optimal payloads, more suitable types and sizes of bits, down-hole motors, instruments, casing and drill pipe, blow-out preventers and well-head equipment.

In the Tatar republic, where more than 1,000 wells of total depth 1.7 mn metres are drilled each year, better well design has helped reduce the well construction period, enabling an extra 500,000 tons of oil a year to be obtained. The use of the correct types of bit can extend bit runs by 25 to 30 per cent, according to the Drilling Technology Institue, VNIIBT. Accordingly, the Ministry of Chemical Engineering has been setting up a centralised system of bit supply. It was first established in the Tatar republic in 1970, and later extended to the Bashkir and Kuibyshev regions.

Secondly, technological improvements have not been made available to all regions. Many crews still work with obsolete rigs and poor-quality bits, down-hole motors, muds, etc. Thirdly, the quality of organisation varies widely, particularly in spheres such as the optimisation of round-trip-operations and lowering casing pipe, flushing and cementing, and carrying out measures for overcoming breakdowns (fishing, etc.). The state of transport and information services varies widely, even though all associations are now equipped with Minsk-22 and Minsk-32 computers, and all oil and gas production directorates have Burroughs TS-3500 terminals.

For oilfield management, the quality of organisation largely depends on the extent of automation. The Eleventh FYP has called for the operation of 80 to 85 per cent of the USSR's oil wells to be automated by 1985, and some long-established fields like that of the Bashkir republic are already 98 per cent automated. Almost 13,000 wells at 41 out of 46 Bashkir deposits are under electronic control, and this has resulted in large savings in servicing costs. During the 1970s, the number of Bashkir wells almost doubled but the number of servicing personnel hardly increased. This has meant that service workers have had to become much more versatile - the Chekmagushneft directorate, for example, has crews of 4 workers, each with several professions, servicing 60-70 wells per crew. The effect of automation, and the introduction of the brigade method of labour organisation in the Bashkir republic, has allowed 9,000 workers to be re-allocated, and a single-shift system has been adopted, leading to a fall in labour turnover. The centrepiece of the automation system is the "Sputnik" measuring facility which collects daily data on all aspects of the operation of each well and presents it for processing by the Minsk-32 computer. The "Sputnik" was designed by the "Soyuzavtomatika" Association of Oktyabrskii whose automation systems are employed at 20,000 wells in the Bashkir and Tatar republics and Western Siberia.

The increasing incidence of cluster drilling is permitting some organisational problems to be eased. This has taken place mainly in Western Siberia, but the practice has been extended to other regions. There are 24 artificial islands for cluster drilling in the Nizhnekamsk Reservoir, for example. In 1975, 60.5 per cent of all Soviet development drilling was inclined compared with 10.7 per cent in 1965.

In 1970, there were 50 clusters of 157 wells in Western Siberia, and by 1975 there were 216 clusters of 1,089 wells. The plan called for 780 clusters by 1980. The number of wells per island reaches a maximum of 24 at Samotlor, but in the Caspian Sea up to 34 wells have been drilled from one island at Neftyaniye Kamny. The greatest angle of inclination is 45 degrees, with deposits 2,500 metres deep being tapped at a distance 2,500 metres from the well-head on a horizontal plane.

As well as facilitating the organisation of transport and communications facil-ities, inclined drilling has technical advantages which allow drilling speeds to be raised. In 1975, the commercial speed of drilling inclined wells in Western Siberia averaged 3,904 metres per rig per month compared with 3,739 for all wells. The installation of cluster pumping stations permits great cost savings to be made. During 1971-1975, 91 such clusters were installed in Siberia, and the plan for 1976-1980 was to build 182.

The Moscow Research Institute for Drilling Technology has also designed equipment for drilling horizontal bores over great distances, and it was first used by Almetyevsk (Tatar republic) drillers. Some "tens" of wells, each with five to ten horizontal bores, are said to have been drilled.

Waterflooding has been used extensively in the USSR since it was first employed at Tuimazy (Bashkir republic) in 1946. It has been the subject of a considerable amount of debate among oil experts, with the Soviet practice of contour flooding attracting particular criticism from Western observers.

Soviet oilfields are flooded from the very beginning of their development. Consequently, 89 per cent of all oil produced is extracted with the aid of secondary recovery, mostly waterflooding, at 260 deposits. In Western Siberia, 99 per cent of all oil is obtained with waterflooding.

Table 27

Waterflooding

Oil obtained by waterflooding	1965	1970	1975	1979	1980
(% of total)	70	74	86	87	89
Water injected (mn cu m)	329	559	985	1,407	1,559
Water per ton of oil (cu m)	1.93	2.14	2.34	2.76	2.90
Water cut (%)	41.2	43.9	48.2	52.7	54.5
Fluid lifted (mn tons)	289	466	815	1,077	1,179

Sources: Narodnoe khozyaistvo 1980, Ekonomika Neftyanoi Promyshlennosti & Neftyanoe Khozyaistvo (various issues)

The purpose of waterflooding is twofold: to maintain the pressure in the reservoir, which enables the well to flow freely for a longer period; and to raise the recovery rate by scouring oil from the reservoir. Critics of the policy argue that under the Soviet practice of contour flooding, water soon breaks through to the bottom of the oil-well, and higher initial yields are eventually accompanied by higher water-cuts and lower recovery rates.

Perimeter flooding, whereby water-injection wells are drilled at the edge of a deposit, and the water forces the oil towards the oil-wells in its centre, is a widely established and highly successful practice in the West. Contour flooding is where water-injection wells are drilled throughout the deposit in different types of networks, dividing up into different sections, each of which can be worked independently. The method can be highly successful if all the characteristics of the deposit are known - its size and structure, the nature of the reservoir rock (porosity, permeability), the nature of the oil (its gravity, viscosity, gas content, migration characteristics) and other factors such as the nature and volume of reservoir water etc. It has been argued by Soviet experts that when contour flooding was first carried out at Romashkino in 1949, these characteristics were not fully understood. During the 1950s, the new method was extensively used, both at new deposits and at old ones previously subjected to perimeter flooding.

The question of waterflooding is intimately connected with that of well location, and contour flooding was initially accompanied by sparse networks of development wells, usually 30 to 36 ha per well, and in some cases 48 to 60.

The regions with the most favourable characteristics for contour flooding include Western Siberia and the Turkmen republic. At the Bashkir and Udmurt republics, Perm oblast and the Mangyshlak peninsula, various problems such as high viscosity oil, oil with high paraffin content, and complex reservoir structures made the method developed at Romashkino less effective and it became necessary to develop new systems of waterflooding and well location more applicable to diverse geological and physical characteristics of the deposit.

In October 1973, a conference of oilworkers at Almetyevsk approved the introduction of a new type of waterflooding, pattern flooding, whereby deposits are sectionalised into narrow bands with a three-row distribution of oil wells following the migration routes of the oil. This would prove particularly useful for the exploitation of pools in terrigenous and carbonate reservoirs where wells have low yields and the oil has a relative viscosity of up to 30 centipoise. Pattern flooding enables the location of water injection and oil wells to correspond more closely to the geological and physical characteristics of the deposit.

Russian experts claim that during the three decades they have been used, water-flood systems have been constantly modified and improved, as the characteristics of oil deposits have become better understood. While Western observers claim that the ultimate recovery rate is reduced by Soviet waterflooding practices, the Russians say that, on the contrary, it is increased.

Waterflooded reservoirs in the Soviet Union have an average recovery rate of 46 per cent, compared with average American rates of only 32 to 33 per cent, and the task now is to raise them to 55-65 per cent. This will be achieved by more intensive (and better planned) waterflooding and with the use of tertiary recovery methods.

Towards the end of 1977, the recovery rate at the D_1 payzone of the Tuimazy deposit had reached 52 per cent, at Bavlinsk D_1 it was 52 per cent, at Serafimo-Leonidovsk D_0+D_1 it was 49 per cent and at Shkapovo D_4 it was 45 per cent. A year later, the three Bashkir deposits of Tuimazy, Shkapovo and Serafimo-Leonidovsk between them had a water-cut of 85 per cent, the recovery rate had reached 48 per cent and they were still producing oil at a rate of 12 mn tons/year. It is claimed that, without waterflooding, these deposits would have become exhausted with no more than 30 to 35 per cent of the oil extracted. Moreover, the capital and running costs would have been many times greater, and the duration of exploration would have been much longer.

Some Western observers have criticised Soviet waterflooding practices without really understanding how they work. It was even once thought that the water-cut might rise regularly by six percentage points a year from 50 per cent in 1975 to 80 per cent in 1980, thereby presenting the Soviets with a fluid-lifting requirement in that year of 3,200 million tons compared with 815 mn tons in 1975.

The fact is that the water-cut does not necessarily rise regularly for the country as a whole. It rises or falls depending on the rate at which new productive capacity is brought on-stream, and old wells are shut down. Even for individual wells, the water-cut does not rise regularly - it can rise rapidly, then slowly, and sometimes it even falls, depending on formation pressure changes and changes in the direction of fluid movement in the reservoir. But eventually it will rise to 97 to 98 per cent, at which point the cost of pumping will exceed the value of the oil being obtained, and the well will be shut down.

Nor does the water-cut for the whole deposit rise regularly. Soviet research shows that it grows most rapidly when the recovery rate lies between 30 and 45 per cent. The biggest annual increases in the water-cut at Bashkir fields were 14 per cent at Shkapovo D4 in 1969 and 9.3 per cent in Tuimazy D1 in 1968, but in recent years, with the recovery rate approaching 50 per cent, the water-cuts have grown by only 1.4 to 1.8 per cent a year.

Some deposits can operate for many years with water-cuts of 80 per cent or even 90 per cent depending on the value of the oil produced and the cost of water injection and fluid pumping. The cost of producing a ton of oil rises fivefold when the water-cut at the deposit increases from 50 per cent to 90 per cent.

For the USSR as a whole, the Tenth FYP foresaw the water-cut rising from 48.2 per cent in 1975 to 57 per cent in 1980. In fact it reached only 54.5 per cent, compared with my own prediction of 53 to 54 per cent[1] and a CIA forecast of 65 to 80 per cent[2]; it can be shown that the latter forecast was technically impossible.

The water-cut has grown much more slowly than the CIA thought possible, not only in the heavily flooded Urals-Volga region but also in Western Siberia. In 1975, the water-cut at the Nizhnevartovsk field varied between 25 per cent at Megion after eleven years of production and 2.5 per cent at Vatinsk after nine years. At Samotlor it varied considerably over the twelve payzones, reaching a maximum of 14 per cent. The major deposits in the Surgut field had very low water-cuts; 2.5 per cent at Mamontovo after five years and 5 per cent at Pravdinsk after seven years. Only at the early deposits at Zapadno-Surgut and Ust Balyk (both peaked in 1972) has it reached 30 per cent and new payzones are being exploited which may reduce it again.

Improvements in waterflooding methods

It has already been mentioned that improvements to various aspects of waterflooding are constantly being carried out, and since the Almetyevsk conference of 1973, the Ministry of Oil has periodically organised similar conferences. These help to disseminate the results of the latest research into new technology and new methods of exploitation of deposits. During the periods between these conferences, a Central Commission for the Exploitation of Oil Deposits operates. It consists of scientists and oilmen from all the producing regions and considers and selects projects for the exploitation of individual oil deposits.

But in spite of this work, it was noted in early 1979 that methods for raising the recovery rate had not yet gained widespread application, according to a speaker at a Tyumen oblast conference on the development of the oil and gasfields.

The problem of water breakthrough is most acute in the Tatar republic. Each year, more than 600 wells have to be repaired to reduce the water flow, and two special directorates have been set up at Almetyevsk and Leninogorsk, charged with the task of raising recovery rates and carrying out well repairs. During the four years 1973 to 1976, they insulated 1,589 wells from breakthroughs by bottom, injection, surface and ground water. These repairs, which cost 21 mn roubles, reduced the annual water flow by 14.5 mn cu m, and enabled an additional 9.2 mn tons of oil to be extracted. The cost of such a well repair, estimated at up to 10,000 roubles, is recouped within an average of 20.6 months, and the average duration of the effect of the repair amounts to 25.9 months. However, a large number of repairs have only a short term effect for other reasons, and when these are excluded from the calculation, the average duration of the effect rises to 40.6 months.

---

1 EIU "Soviet Oil and Gas to 1990", p. 78   2 CIA July 1977, p. 17

In all, the volume of water which was lifted with each additional ton of oil was reduced by 1.57 cu m, and the water-cut for the repaired wells was reduced by an average 39 per cent.

There is said to be a vast potential for improvement, particularly from the correct choice of insulation materials. Six types of material are used in the Tatar republic, and extensive tests have been carried out to compare their effectiveness in plugging breakthroughs by different types of water. It has been confirmed that cement is best for plugging injection water while cement-resin compounds were slightly better for bottom and surface water.

The proportion of successful insulation operations is considered to be far too low, ranging from 58.6 per cent in the Tatar republic to 30 per cent in Kuiby-shev, and averaging 54 per cent for the USSR. Insulation work is usually carried out in wells producing from several seams, and especially where low-productivity seams with a low water-cut are precluded from operating by the existence of high-productivity seams where the oil has been exhausted and which are producing pure water.

Nearly all water insulation repairs in the USSR are carried out with the use of cement, and it is apparent that savings can be made if other materials are used where applicable.

The problem of water treatment is most acute in Western Siberia, where surface water now accounts for nearly three quarters of all injected water.

Table 28

Type of Water Injected into Reservoirs in Western Siberia

	1970	1975	1978
Volume of water (mn cu m)	67.3	259.2	...
of which:			
surface water (%)	28.7	71.4	73.6
ground water (%)	71.1	21.3	12.9
associated water (%)	0.2	7.3	13.5

Sources: Muravienko and Kremneva p. 73, ENP 1979 No. 6

The sharp change in the share of ground water from 71 per cent in 1970 to 12.9 per cent in 1978 is due to the rise of Samotlor where the Senomanskii water-bearing seam is less productive and where large volumes of surface water have been used.

Ground water is of the best quality for waterflooding purposes and therefore exhibits better properties for flushing and forcing out oil than surface water. It also has been filtrational properties, but its use is declining at the three large deposits of Ust Balyk, Vatinsk and Zapadno-Surgut where it is most used. This is because of declining formation pressures in the Senomanskii strata and a shortage of pumping equipment.

The use of associated water has been growing in Siberia; during the period 1970-1978, 50 mn cu m of associated water were pumped, and it is planned that by 1980 it will account for 18 per cent of total pumped water compared with 7.3 per cent in 1975.

Associated water is the cheapest type to use because it does not require the construction of expensive water supply systems. At Samotlor, it costs 35 kopecks per cu m compared with 37 for surface water. While ground water is not used at Samotlor, its cost at other deposits ranges from 59 kopecks at Vatinsk to 65 at Ust Balyk. The use of associated water is estimated to have saved 20.5 mn roubles of capital investment so far.

The basic method of extracting underground water in Siberia at the present time is the UETsP-16-3000-1000 electric submersible pump. Since 1970, these pumps (of which there are now 145) have pumped 110 mn cu m, or 30 per cent of all utilised underground water. They prove much cheaper to use than other pumps like the ETsV-10-120-60 electric pump and the NSV-15 jet pump.

In recent years, other electric pumps such as the UETsP-14-5000-60, the UETsN-56-1000-700 and the UETsN-6-1000-650 have been tested in water-supply wells in an inverted position so as to force water from aquifers into oil-bearing strata. The most effective pump is the UETsP-14-5000-60, which pumps water 30 per cent more cheaply than established units.

The problem with surface water is that it has to be cleaned before use, as otherwise suspended particles and chemical impurities block the pores in the reservoir rock. Cleaning is accomplished by filtration, costing 8 kopecks per cu m, chemical methods using reagents and coagulants (11.6 k/cu m) and electro-coagulation (30.1 k/cu m). The use of a filtration plant of 7,000 cu m capacity a day gives a saving of 5.4 mn roubles a year over chemical methods.

It is claimed that great successes have been achieved in oilfield geophysical work, especially in the study of the collector properties of reservoirs and of the migration characteristics of oil and water.

Cyclical waterflooding has been carried out in the Tatar republic and Western Siberia. It generally consists of a gradual increase in the volume of water pumped during a period of 10 to 30 days, then a gradual decline for a similar or longer period, up to one year at old deposits. The volume of water pumped is usually reduced to 60-90 per cent of normal, and the growth in the water-cut is decelerated without loss of oil.

At Samotlor, cyclical waterflooding has been employed since 1975 in the south-east and north-west sections of the AV_{2+3} payzone. The water volume is increased for 30 days, then decreased for 30 days. From 29 wells, 365,000 tons of additional oil have been obtained for only 83 per cent the normal volume of water. From the 14 Tyumen deposits where cyclical flooding has been tried, an additional 677,000 tons of oil have been obtained, and 2.54 mn cu m of water saved.

WATERFLOODING AND WELL DENSITY

The growth of waterflooding was accompanied by a considerable thinning out of the well network. The higher yields of wells in fields subjected to flooding meant that a stable rate of extraction of oil could be achieved with a smaller number of wells than before. This in turn led to an increase of six to ten times in the level of reserves per well. However, when the deposit begins to age, unit costs of oil extraction start to rise rapidly due to the increased share of costs that have to be spent on maintaining the level of extraction. This has been true of most large Urals-Volga fields during the last decade.

A fierce controversy has arisen over the question of oilfield development in
relation to waterflooding. Some experts, such as B.T. Baishev, maintain
that until the introduction of waterflooding, the exploitation of oil reserves
took place at very low rates of 1 to 2 per cent of recoverable reserves a year.
A principal advantage of waterflooding, according to Baishev, was that it
enabled these rates to be increased greatly. Others, like A.P.Krylov,
chairman of the Scientific Council for Problems of Exploration of Oil Deposits,
argue that the precise opposite is the case[1]. He says that, until the advent
of waterflooding, exploitation rates for the sector were high, but fell fourfold
during the first ten to fifteen years of waterflooding.

Krylov claims that arguments in favour of sparser well networks were advanced
long ago, but only with the introduction of waterflooding was it possible to
overcome the barrier of conservatism and achieve an appreciable thinning out of
the well network. Thus, although individual well-yields were raised, the rate
of exploitation of a deposit as a whole declined, and this enabled deposits
to be worked for longer periods of time.

But since 1968, the rate at which output from the same well declined from year
to year has grown steadily. This means that old wells are providing a progressively smaller share of total output, with new wells giving a proportionately
larger share. Thus the number of new wells being drilled each year must continuously increase, simply to maintain output. Krylov goes on to assert that
"preliminary calculations show that if we maintain the current rate of annual
increase in the number of new development wells, and the rate of decline of
output from old wells continues to grow, then within a comparatively short period,
the extraction of oil in the country will achieve its maximum, after which it
will begin to fall".

It has been suggested that Krylov was confirming a prediction by some Western
observers that Soviet oil production was about to decline. However, he goes on
to say, "To change this course of events, and reach the planned volume of extraction, is possible either by increasing the rate of growth of new wells (requiring increased capital investment and expenditure of pipe) or by transferring
to an economically and technologically well-founded system of exploitation leading
to a thinning out of the well network and a reduction in the rate of decline
of output from old wells (not requiring additional capital investment)."

Krylov is merely saying that if the USSR wants to get more out of the oil
industry it must either put more into it, or organise it better. It should also
be pointed out that Krylov may be somewhat pessimistic. His analysis is undoubtedly
true for the Urals-Volga region; other specialists such Guzhnovskii[2] confirm that
the rate of fall of output from old wells is rising rapidly, but argue against trying
to counteract this by infill drilling. Guzhnovskii says this is inefficient and
wasteful, and that the resources thus spent could be better utilised at new deposits
in Western Siberia. Whether Krylov's analysis is correct for Western Siberia is
less obvious. Some people allege that Samotlor has been overworked while other
deposits have been neglected. But this argument is becoming increasingly
difficult to sustain; it implies a rapid decline in output from Samotlor, whereas production has been maintained at 150 to 155 mn tons a year, and the Soviets
expect it to remain at this level until beyond 1985 at least.

According to Krylov, the increase in the well density, and hence the fall in
the volume of reserves per well, has caused the rate of decline of output
per well to accelerate by 150 per cent during the last decade, while average

---

1 EKO 1980 No. 1   2 EKO 1979 No. 2

well-yields have risen by only 25 per cent. It is argued that the increase in the well density is the main reason for the problems which Krylov believes may arise.

Increased well densities occur for one of three reasons.

1. The failure to prepare reserves sufficiently quickly. This, assuming an unchanged reserves-per-well ratio, is said to lead to an increase in the ratio of category A reserves to total industrial reserves. In fact, this ratio has actually been reduced and increased well densities do not stem from a shortage of reserves.

2. Technological considerations, i.e. the achievement of the projected recovery rate and the newly adopted criteria of optimality of exploitation. Krylov believes that the projected recovery rate is estimated on the basis of a comparatively small volume of data, and can be estimated only approximately. Consequently, an unnecessarily dense network may be drilled in an attempt to achieve an unjustifiably high recovery rate target.

    Research has shown that under waterflood conditions, the recovery rate is a product of the two coefficients of displacement and conformance, and if the water-flooding operating is performed correctly, then it should be possible to reduce the well density without affecting the eventual recovery rate. This has been confirmed by several groups of researchers and in 1957 an experiment was carried out at Bavlinsk deposit where the oil has a comparatively low viscosity of up to 5-6 centipoise.

    The experiment consisted of plugging half the wells in the main part of the deposit (drilled to 20 ha per well) which effectively gave it a network of 40 ha per well. The deposit was flooded, and after the water had broken through, the plugged wells were reopened. This made it possible to estimate the effects of doubling the well density. In fact, it is apparent that the recovery rate can be maximised, not by an arbitrary condensation of the well network, but by the drilling of a "reserve fund" of wells, amounting to 10 to 30 per cent of the basic fund, in the areas where the basic wells cannot remove all the oil. An arbitrary redrilling of the deposit is said to have an insignificant impact on recovery rates.

    Krylov accuses the 1973 Almetyevsk conference of the Ministry of Oil of adopting revised criteria for the exploitation of oil deposits which encourage greater well densities. Fortunately, far more work is now being carried out on the elaboration of optimal schemes of exploitation of deposits.

3. The third factor leading to greater well densities is the lack of manoeuvrability of drillers. While this may be due to the lack of roads in some areas, it is primarily an organisational problem. Each association employs a certain number of drillers, which have to be employed even if the association is working a declining field. Therefore they are set to work on drilling more wells, and increasing the well network density. Krylov proposes that drillers should work independently of the oil producing associations, so that drilling can be concentrated where it is most needed and not where a particular association employs most drillers. It can be argued that the current practice of associations sending drilling crews to work in Western Siberia on a tour-of-duty basis helps to meet Krylov's proposal without the necessity of introducing wide ranging organisational changes.

Krylov's article was an interesting and timely contribution to the research into oilfield management taking place in the USSR. It is quite wrong to say that he "has predicted that national oil output will peak in a relatively short time and then start to fall".[1] He has simply pointed out what he perceives to be the danger of unjustifiably dense well networks, and has argued that either this practice should be changed, or considerable increases in drilling capacity will be required in the future. The Russians, as always, appear to be compromising between these two policies by providing for vastly increased volumes of drilling in 1985 (see Chapter 10) and by adopting thinner well networks at new fields in Siberia.

In the past, wells have been drilled in networks the density of which depends on anticipated well yields. For low well yields, the density has been 16, 25 or 36 ha per well (i.e. wells regularly spaced 400, 500 or 600 metres apart). Average-yield wells were located at 25, 36 or 49 ha per well, and high-yield wells at 25, 36, 49, 64 or 81 ha per well. Most new Siberian fields are being drilled with three rows of development wells and a row of injection wells at a distance of 800 to 1,200 metres. Well spacing is comparatively thin at 36 to 56 ha per well, and the latest fields are being drilled at 81 to 100 ha per well. The trend towards thinner well networks is continuing, but is limited by the practice of cluster drilling. With an angle of inclination not exceeding 45 degrees, and a depth of 2,500 metres, it is possible to drill 25 wells 800 metres apart (64 ha per well) from one island, and this has become standard for deposits like Samotlor. By thinning the well density, the number of wells that can be drilled from one island falls, and their unit cost rises because fixed costs like rig transportation and assembly have to be spread over a smaller number of wells.

TERTIARY RECOVERY METHODS

During the four years 1976-1979, experimental work involving the use of different tertiary recovery methods took place at 45 deposits, mainly in the Tatar and Bashkir republics and Kuibyshev oblast. These experiments appear to have achieved their limited objectives, with 8 mn tons of extra oil being obtained. Three mn tons came from the Tatar republic, and the Eleventh FYP has set a target for the republic of 6 mn tons over 1981-1985, or 2% of total output.

Some methods are now being introduced on a large scale basis (waterflooding with solvents of surfactants and polymers, the pumping of caustic soda into the reservoir, the use of sulphuric acid, sodium triphosphate and foaming agents), while others are being tested under industrial conditions (waterflooding with miscible solvents, the pumping of liquid carbon dioxide, etc.). However, the most important oil-producing regions are failing to make full use of these new methods due to a shortage of materials and a poor organisation of work, and the responsible associations, Soyuztermneft and Soyuznefteprom-khim, have been severely criticised.

Experiments using water mixed with chemical reagents have been carried out in 75 places at 22 deposits over a number of years. A solution of water and polyacrylamide has been pumped into Orlansk deposit (Kuibyshev) since 1966, but the results have proved inconclusive. Since 1977, the same experiment has been carried out at Sosnovsk deposit belonging to the Kuibyshevneft Association. Some 98,700 cu m of solution was pumped into the deposit at a

---

1 _Time_ 23/6/80

capital cost of 197,800 roubles. A small amount of additional oil was obtained, and when the value of this oil was compared with the capital cost plus additional running costs of producing the extra oil, an annual increase in profits of 300,000 roubles was obtained. However, the effort and expense in acquiring this additional oil at Sosnovsk reduced the resources of the association for oil extraction at other deposits. The money could have been spent on simply drilling new production wells, and consequently the additional profit for the association was estimated at only 76,300 roubles.

It is calculated that 1 ton of polyacrylamide (at 100 per cent concentration) and a lot of effort is needed to obtain 200 to 300 tons of oil, but in spite of these small returns, experiments are continuing, and in the long term the use of these solutions may prove more effective.

The experimental pumping of alkaline solutions has taken place in Western Siberia, Azerbaidzhan and Perm oblast, giving 20 to 150 tons of additional oil per ton of 100% reagent, and raising the recovery rate by 2 to 15 per cent. At Trekhozernyi deposit (Western Siberia) the periodic pumping of a silicate alkaline solution is expected to raise the recovery rate by 5 per cent compared with normal waterflooding, with 48 tons of additional oil per ton of reagent anticipated.

Interest in using liquified carbon dioxide has been stimulated by research by the Bashkir Oil Research Institute, which shows that it can permit recovery rates to be raised by 6 to 20 per cent compared with 4 to 10 per cent by using surfactants.

The large-scale testing of methods involving liquid carbon dioxide has started at the Bitkovsk, Zhuravlevsk-Stepanovsk and Romashkino deposits. The addition of carbon dioxide to water being pumped into the reservoir considerably increases the capillary displacement of oil and the phase permeability of the reservoir. It has been established that a 4 to 5 per cent gas content in water injected from the beginning of a reservoir's exploitation will increase the recovery rate by 10 to 15 per cent. Different variants of using carbon dioxide have been tested: continuous pumping, cyclical pumping, and the simultaneous pumping of carbon dioxide gas and water. The use of carbon dioxide is particularly suitable for deposits with heavy, high-viscosity oil.

The marginal cost of oil obtained by the use of surfactants and carbon dioxide is very high. For surfactants, it ranges between 40 and 80 roubles per ton everywhere except the Tatar republic and Tyumen oblast, where it can fall to as little as 20 roubles a ton. For liquified carbon dioxide, it varies between 40 and 80 roubles a ton in the Bashkir, Kuibyshev and Mangyshlak regions, 60 to 80 roubles a ton at Perm and Belorussia and 80 roubles a ton in the Tatar republic. It has not yet been used in the Komi, Azerbaidzhan or Tyumen regions. The new control prices for oil should make the use of both surfactants and carbon dioxide much more profitable.

During recent years, there has been an increase in the range and volume of tertiary agents supplied by the main producers such as the Sterlitamak petro-chemical plant. Much of the new technology is being imported, and recent purchases include a plant for producing 250,000 tons of surfactants a year from Pressindustria of Italy costing $24.5 mn, a carbon dioxide liquifaction plant of 400,000 tons capacity from Borsig of West Germany to be installed near Kemerovo, and a similar plant of 1 mn tons capacity. In August 1980, the French firms Technip and Entrepose were awarded contracts worth $100 mn for the supply of injection equipment to be used at the Romashkino field. The liquified carbon dioxide is to be supplied from the Tolyatti plant through a 250 km pipeline.

So far, the use of waterflooding with hot water or steam has taken place only as industrial tests at 28 places in 19 deposits, but full scale introduction is now due. This is especially necessary where waterflooding lowers formation temperatures significantly, for example at Romashkino where they are reduced from 34 to 10-20 degrees in some wells, thereby leading to paraffin deposition. The most important steam injection experiments have been carried out at Kenkiyak in the Emba field of Kazakhstan and at the Yarega and Usa deposits in the Komi republic. The use of steaming in conjunction with cold water flooding at the Okha deposit (Sakhalin) has yielded 1.36 mn tons for a saving of 32.3 mn roubles.

At Yarega, steam injection can raise the recovery rate to 40-50 per cent in some sections, compared with only 4 per cent under normal working. Steam is being pumped at a rate of 20 tons an hour, and this should raise oil output by 0.5 mn tons a year. While research at Kenkiyak has been taking place for a longer period of time, that at Usa is more important because this is a super-giant deposit which is currently producing 8 mn tons a year of mostly heavy oil. Research has shown that oil yields can be raised four to five times, which would justify the use of steam injection in spite of the high costs involved.

These would increase the capital cost of developing the field of 280 wells from 147 mn to 262 mn roubles. The additional cost would consist mostly of drilling steam injection wells, building boilers, providing the water supply and changing the oil preparation installations. Steam injection would raise the annual output of oil from 8.6 to 38 mn tons, which could only be obtained under normal circumstances by the drilling of 968 additional wells costing a further 503 mn roubles. Moreover, exploration costs would rise from 21.1 to 93 mn roubles if production were to be increased to 38 mn tons/year by normal methods.

Thus it can be shown that steam injection can provide savings of 387 mn roubles in capital costs and 72 mn roubles in exploration costs, and this method will be introduced extensively at heavy oil deposits during the 1980s, especially after the new steam injection equipment plant, currently under construction at Taganrog, comes on-stream. The first of these deposits is Gremikhinsk in the Udmurt republic, which began producing in 1981. The steam is pumped at 120 to 160 ats and a temperature of 340 degrees to a depth of 1,189 metres.

Experiments with pumping hot water have been carried out at Arlan (Bashkir republic) and Uzen (Mangyshlak). At Arlan, specific injectivity increased by 35 per cent and reservoir conformance by 15-30 per cent compared with cold water flooding. The total increase in the recovery rate amounted to 13.4 per cent.

In situ combustion has been carried out at the Pavlova Gora, Khorasany, Koshanaur and Severnaya Skladka deposits in Azerbaidzhan. Yields 20 to 30 per cent higher than with waterflooding were achieved. At the Buzachi deposit of Karazhanbas, in situ combustion is being carried out in conjunction with steaming by Soyuztermneft, and a new combustion facility is being tested by the Malgobe-kneft directorate of Grozneft in the Checheno-Ingush republic.

In order to encourage the greater use of tertiary recovery, the Ministry of Oil has set up a special incentive fund from which the country's 26 oil producing associations will receive a payment for each ton of oil recovered with the use of tertiary recovery.

Secondary and tertiary recovery problems are most apparent in the regions of declining production, especially the Urals-Volga region. This region not only produces nearly 200 mn tons of oil a year, but has an extensive transportation

110

network and contains nearly half of the USSR's refining capacity. The opportunity cost of each additional ton of oil is very high, and the next decade should see strenuous efforts to arrest or slow down the decline in output. A similar situation is to be found in the Caucasus and North Caucasus regions.

Problems in other regions are of a quite different nature. During the next ten years, Western Siberia will account for the whole of the increase in Soviet oil and gas production as well as compensate for most of the decline in the older producing regions. In this it will be assisted to a small extent by the Caspian Sea regions (for oil) and Central Asia (for gas). The following sections will examine the problems specific to these regions, because the USSR's ability to produce increasing volumes of oil and gas will depend on its ability to overcome these problems.

CHAPTER 8

Problems of Infrastructure in Western Siberia

POPULATION

Twenty years ago, the USSR's principal oil-producing region, the Khanti-Mansiiskii national okrug, and its main gas-producing region, the Yamal-Nenetskii national okrug, both part of Tyumen oblast, were an empty wilderness, inhabited only by indigenous reindeer herdsmen and the workers of a small timber industry. Nizhnevartovsk did not exist and Surgut was a tiny river port handling fish, timber and furs. In short, the entire infrastructure had to be built up from scratch. In order to attract people into the region, whole towns had to be built. The standard of the housing and the quality of the amenities had to be far superior to those of the rest of the country in order to attract people to a region where temperatures average -50 degrees in winter, and where swarms of stinging insects thrive during the short but hot summer. The high winds, perpetual snow-haze and very long nights in winter make the region singularly unattractive for permanent residence.

At first, the Russians believed that they could attract sufficient workers simply by paying large bonuses, but the shortage of work for women in the region meant that the average income per family was not much higher in Tyumen than it was in Moscow. As the infrastructures of Nizhnevartovsk and Surgut develop, more and more work is becoming available for women, but many oil and gas workers are still single men who come to the region to work for two years, earn sufficient money for a flat or a car, and then return to "the mainland" as the rest of the USSR is called. There are signs, however, that the population is becoming increasingly settled with the labour turnover rate falling to acceptable levels.

This is reflected in an accelerating rate of growth of population. The same number of people are migrating into the region as before, but more of them are finding it worthwhile to stay. Both Nizhnevartovsk and Surgut now have populations large enough to justify a wide range of amenities, and are rapidly losing their frontier outpost appearance. Labour turnover is down to 20-22 per cent, the same as in the Urals region, although sociologists consider 8 to 10 per cent to be optimal.

In 1959, Tyumen oblast had only 1,092,000 inhabitants, with only 186,000 living in the Khanti-Mansiiskii and Yamal-Nenetskii regions. By 1981, the oblasts population had grown to 2,031,000, including 865,000 living in the two sub-divisions. All the 679,000 new people living in the Khanti-Mansiiskii and Yamal-Nenetskii regions are involved directly or indirectly in the oil and gas industry, there being no other industrial activity other than fish-canning and timber processing.

Table 29

The Population of Tyumen Oblast
('000, beginning of year)

	1959	1970	1975	1979	1980	1981
Tyumen oblast	1,092	1,407	1,580	1,887	1,954	2,031
of which:						
Khanti-Mansiiskii	124	272	390	569	616	672
Yamal-Nenetskii	62	80	118	158	176	193
rest of Tyumen oblast	906	1,055	1,072	1,160	1,162	1,166
Cities:						
Tyumen	150	269	323	359	369	378
Nizhnevartovsk	-	16	52	109	122	134
Surgut	6	34	60	107	121	137
Tobolsk	36	49	49	62	64	65
Nefteyugansk	-	20	n.a.	65	51	57

Source: Narodnoe Khozyaistvo

The significant features of Table 29 are that the striking growth of the Khanti-Mansiiskii region is accounted for largely by the two major cities of Surgut and Nizhnevartovsk (which both had more than 150,000 inhabitants by the beginning of 1982) and that the population of the rest of Tyumen oblast is continuing to grow in spite of heavy migration out of the region from rural areas. This is partly due to the hierarchical structure of settlements adopted as the strategy for developing the region.

Thre are four levels of settlement. The first consists of "the mainland", i.e. the rest of Tyumen oblast dominated by Tyumen City, and some cities and regions beyond the Tyumen border. These are assigned the tasks of providing technical and engineering support, delivering agricultural produce to the oil and gas fields, and fulfilling the educational and scientific requirements. The second level consists of the base cities of Nizhnevartovsk, Surgut, Urai and Strezhevoi for the oilfields, Nadym and Novyi Urengoi for the gasfields and Tobolsk for the petrochemical sector. The smaller cities and settlements (Nefteyugansk, Megion etc.) constitute a third level of outposts, and the fourth level consists of tour-of-duty settlements with few amenities and a constantly changing population.

The rapid improvements in oil and gas technology and the growth of mechanisation and automation are helping to keep down labour requirements. Otherwise, 400,000 workers would be needed by Glavtyumenneftegaz alone in 1985 to meet its plan targets, compared with 90,000 in 1978, but this number can be reduced to about 250,000. The big increase in the volume of geological work in Tyumen oblast called for by the 11th FYP will require a further 100,000 workers if there is no rise in labour productivity. Even if the planned rise in labour productivity of 30 to 35 per cent is achieved, the number of geologists must double over 1981-1985.

Between 1980 and 1983, the number of workers in the oil and gas production and geological surveying sectors is planned to rise by 200,000, and the number of workers in the servicing sector will also rise considerably. Overall, the number of workers in the oil and gas regions is expected to double by 1985 over 1978. This would give a population at the beginning of 1986 of 1,450,000 and an increase of 107,000 a year compared with 59,000 a year during 1975-1981.

In the "perspective", which is believed to mean 1990, the population is expected to reach 2 million, giving an increase of 110,000 a year over 1986-1990.

Most of these new inhabitants will live in the three major cities of Surgut, Nizhnevartovsk and Nefteyugansk. Surgut is planned to have a population of 175,000 in 1985 and 300,000 by the end of the century, and at Nizhnevartovsk it is planned to bring the population of its projected maximum of 250,000 by as early as 1984. The third largest oilfield town is Nefteyugansk, where growth is inhibited by its location on an island in the middle of a lake. New land for housing has to be acquired by extending the island with sand dredged from the bottom of the lake. Nevertheless, the town is growing three times faster than envisaged by the General Plan published in 1979, which anticipated a population of 100,000 by the end of the century. Current house-building plans suggest that the 100,000 mark should be passed in 1985.

Noyabrskaya, with 15,000 people, will soon become a city and will eventually have 80,000 people. In 1979, the construction of the new city of Khanto, 11 kms north of Noyabrskaya was begun; it will eventually have 70,000 people, and will serve the Kholmogorsk oilfield. The settlement of Kogolymsk, serving the Kogolymsk oilfield 100 kms south of Noyabrskaya, is to see its population grow from 3,000 to 30,000. The settlement of Langepas will soon be elevated to a city, because the discovery of large new pools at the nearby Urevsk oil deposit means that 60,000 people must be housed there in the near future. The planned development of large new oilfields on the Agan river means that the size of Novoagansk, Varyegan and Raduzhnyi must grow rapidly during the period 1981-1985. The main city of the Tomsk oilfield, Strezhevoe, should see its population rise from the current 40,000 to 60,000 by the end of the century, and the largest tour-of-duty settlement is to be Novyi Vasyugan, serving the Vasyugan group of deposits. Its shifting population will rise from the current 40,000 to 60,000 by the end of the century.

Salekhard's role as a supply depot and railhead for the northern gasfields will increase its population from the present 26,000 to 100,000 by 1985. Novyi Urengoi will replace Nadym as the principal centre of the gasfields while growing from 20,000 in 1980 when it became a city, to an eventual 100,000. Nadym, currently with 15,000 people, will grow more slowly to 70,000.

As well as an increase in the population of existing towns and settlements, new settlements will appear on the map. In 1980, work began on 13 new settlements for oil workers, and during the 1981-1985 period, 22 new settlements are to be created, including 15 which will eventually attain city status with populations of 15,000 to 80,000. The new settlements include Kharp, where the first four-storey buildings, schools and shops were built in 1980, Sverdlovskii near Nyakh on the Ivdel-Sergino railway, which will have 25,000 inhabitants and where an engineering base for oil and gas workers is to be set up, and Pudino in Tomsk oblast, where the population is expected to reach 40,000.

The new town of Put Yakh is being built on the Tyumen-Surgut railway to house workers on the Mamontovo oil deposit. By mid-1981, it had more than 3,000 people and is set to grow during the 11th FYP as the development of Mamontovo extends southwards. The town will be administered as a single unit with the nearby settlement of Mamontovo. Another important new population centre will be the town of Nyagan, destined to be the base city for the Krasnoleninsk oil-field which is now being exploited. The decision to build 9-storey blocks of flats suggests that it will eventually have a population of at least 60,000, given the recommended relationship between population size and city structure for the far north.

On the gasfield, one of the most important new settlements will be Tikhii, situated a short distance from the settlement of Urengoi (as distinct from Novyi Urengoi, which lies 80 kms to the west). It is a junction for the Surgut-Urengoi railway line, and will be the site of the planned 4,000 MW Urengoi power station. It is expected to have a population of 150,000 by the end of the century, compared with the "10,000 or so" said to live in the Urengoi/ Tikhii area at the present time.

But in spite of the rapid growth of the permanent population, many more workers are working in Western Siberia on a tour-of-duty basis. They include not only skilled oil workers, but also those involved in building the infrastructure, and 150,000 unskilled Molodogvardeitsy (Young Guards) from the Komsomoi, who have volunteered to build roads, etc.

HOUSING

Perhaps the most crucial problem facing the oil and gas sector is that of the provision of housing for its workers in Western Siberia. It has been estimated, assuming the latest standards of housing per head, that there is a shortage of 4 mn sq m of housing in Tyumen oblast, nearly all of it in the oil and gas regions. In 1978, the deficit was put at 3.6 mn sq m in the oil and gas regions, and over 1976-1980 the shortfall in Nizhnevartovsk alone amounted to 450,000 sq metres. The shortage has been getting worse because of the continuous under-fulfillment of plans for housing construction.

In 1970, 720,000 sq metres of housing were built in Tyumen oblast, and by 1975, this figure had risen by nearly 50 per cent to 1,058,000 sq metres. The Tenth FYP called for 6.5 mn sq m to be built over 1976-1980, and it was argued that at least 2 mn sq m should be built each year by 1980. In spite of the fact that the relatively modest plan for 1976-1980 was significantly underfulfilled with only 5.5 mn sq m being built, the Eleventh FYP has set a target of 11 to 12 mn sq m over 1981-1985, including 6 mn over the control period 1981-1983. This implies that the annual construction rate must rise to 3 mn sq m a year by 1985, or more than twice the figure for 1980.

A long-term plan has been drawn up for the building of 23 to 25 mn sq m of housing in Tyumen oblast during the 10 years 1981-1990, when the population of the oil and gas regions should pass the 2 million mark. On past experience it appears unlikely that these highly ambitious targets will be met.

The root of the problem is that while only one ministry is responsible for the extraction of oil, some 26 organisations belonging to eight different ministries build housing and amenities. For the majority of them, the main responsibility is industrial rather than civil construction, and when they fall behind on plan schedules, or start to run out of money, it is their housing programmes which are cut back.

The Ministry of Construction of Oil and Gas Enterprises builds oil and gasfield installations and pipelines, and is responsible for housing in Surgut, Nefteyugansk, Nadym, Novyi Urengoi, Urai and Khanto. It is one of the best organised ministries, having set up in 1975 a special organisation, Sibzhilstroi, for building houses. It currently has 22,000 Soviet and Bulgarian specialists working for it, and has set up three housebuilding combines for the production of pre-fabricated sections at Tyumen, Surgut and Nadym, plus a factory making wooden houses at Pyshminsk.

The Ministry of Industrial Construction is the other major building ministry. It is responsible for civil construction in the Nizhnevartovsk area, but has failed to meet its plan targets every year since 1974. It was considered necessary to build 3.2 mn sq m of housing in the Nizhnevartovsk area during 1977-1980, but its plan was for only 1.4 mn sq m, and in the event only 1.23 mn sq m (88 per cent of target) were actually built. The construction of amenities is proceeding even more slowly, with the Tenth FYP fulfilled by only 55 to 60 per cent.

Its poor record stems from its failure to build a housebuilding combine in Nizhnevartovsk until 1980. Now that the first stage of the combine has finally come on-stream, impressive plans have been drawn up to increase housebuilding rates from 130,000 sq m in 1979 to 280,000 in 1980, 384,000 in 1981, 414,000 in 1982, and 454,000 in 1983. This would suggest a growth in the permanent population from 109,000 in January 1979 to 246,000 by January 1984. And while some of the new housing will be allotted to workers currently living in temporary housing, it is said that the demand for wooden houses and caravans will also increase.

It should be stressed that the new housing will be utilised not only by workers at nearby oilfields and those involved in creating the local infra-structure, but also by workers employed at the newer oilfields at great distances from Nizhnevartovsk. They will work according to the tour-of-duty method, living in temporary accommodation at their work places with a permanent home in Nizhnevartovsk. The new housing may also be destined for settlements gravitating towards Nizhnevartovsk, such as Megion, Raduzhnyi, Langepas and Pokachevskii, where the rapid growth of oil production is foreseen.

The shortage of permanent housing has led to many workers living in temporary wooden housing and caravans. In 1978, Glavtyumenneftegaz had 90,000 workers, but only 27,000 families lived in their own permanent flats. The rest lived in hostels, caravans, etc. It is recognized that wooden housing can be just as comfortable and cheaper to build than pre-fabricated large-panel blocks of flats, especially in small settlements. Accordingly, 3 mn sq m of wooden housing is to be built during the three years 1981-1983, and for the years beyond 1983 a housebuilding combine is being built in Tyumen. It will produce 500,000 sq m of wooden housing a year. At the moment, several plants build wooden buildings designed for Tyumen oblast. While wooden houses are cheaper, the fastest method of erecting large volumes of housing is by putting up blocks of flats constructed from pre-fabricated sections. Housing construction combines producing these sections already exist at Tyumen, Tomsk, Surgut and Nadym, and the most recent came on-stream at Nizhnevartovsk in 1980. It has an annual capacity of 140,000 sq m, sufficient for 11,000 inhabitants. Until 1980, Minpromstroi railed prefabricated sections in to Nizhnevartovsk from other parts of the country, and it has been stressed that the start of the first stage of the combine will not lead to any falloff in the need to import these sections. On the contrary, deliveries will continue to grow until 1982 when the second stage comes on-stream and will then begin to decline.

Even if the Siberian housebuilding organisations were meeting their plan targets, they would still be unable to keep up with the demand, especially as the Eleventh FYP foresees the population of the oil and gas regions growing at over 100,000 a year. Consequently, organisations from many other parts of the country are joining in. They have pledged to build 1 mn sq m of housing plus social and cultural buildings over the three years 1981-1983.

About 20 per cent of all new housing in Nizhnevartovsk is built by the local branch of Glavmosstroi, the Moscow house-building organisation. Using pre-fabricated panels railed in from Moscow House Building Combine No. 1, Glavmosstroi is responsible for building the higher blocks of flats of 9, 12 and (beginning in 1981) 16 storeys. The panels are given additional insulation and treble-glazing on site.

The Ministry of Construction in the Baltic republics has been given the task of developing the settlement of Kogolymsk, 100 km north of Ortyagun on the Surgat-Urengoi railway, and with a current population of 3,000. By 1983, another 45,000 sq m of housing and 100 kilometres of local roads will have been built by the Latvians and 35,000 sq m by the Lithuanian and Estonian workers who will also be employed. This suggests a population of 10,000 by 1983, and eventually this will rise to 30,000. Housing will be built from pre-fabricated sections produced by the Riga and Vilnius Housebuilding Combines, and other Latvian ministries have undertaken specific tasks in relation to the settlement. The Latvian Ministry of Health, for example, will staff the local health centre.

Another organisation heavily involved in Siberia is the Magnitogorsk Construction trust, belonging to the Ministry of Heavy Industry Construction. Over the four-year control period 1980-1983, it will build 75,000 sq m of housing (sufficient for 6,000 people) at a new settlement for the workers at the Mamontovo oil deposit. The design for the settlement was drawn up by the Chelyabinsk Civil Design Institute.

Leningrad workers are working mainly at Novyi Urengoi, where they are building nine-storey houses in the form of polygons which will each contain 3,000 people. Over the three years 1981-1983, they will build 180,000 sq m of housing in Novyi Urengoi, sufficient for 14,400 people. All these buildings will have walls one metre thick.

A brigade for the Almetyevsk housebuilding combine (Tatar republic) has gone to Surgut, another from Moscow is working in Nizhnevartovsk, and workers from Kazakhstan are helping to build housing in the Tomsk oblast settlements of Novyi Vasyugan and Aleksandrovsk. In addition to housing, a health centre, civic buildings, canteens and warehouses will be erected at Aleksandrovsk by the Kazakh workers. They will also build a goods trans-shipment base for river traffic and five fuel tanks.

Crews from the Ukraine are building 40 blocks of 5 storeys each in Khanto over 1981-1983, and builders from Belorussia and Uzbekistan are working at Strezhevoe and Urai. The Ukrainians are also helping at Nefteyugansk, where 80,000 sq m of housing were built in 1981, and workers from Omsk, Eastern Siberia, the Lower and Central Volga areas, the Tatar republic, and Arkhangelsk are involved in construction projects throughout the oil and gas fields.

Western Siberia's housing requirements would probably prove unattainable if a permanent settlement had to be built at each of the 40 new deposits which will come on-stream during 1981-1985. Most of these, in fact, will be served by tour-of-duty settlements consisting only of housing (mostly hostels) and the barest amenities. The unit cost of housing and infrastructure is twice as great as in the base cities, and 30 per cent greater than in smaller per-manent settlements, and the tour-of-duty method gives cost savings of 16,000 roubles per worker.

The new fields in the Vasyugan swamp of Tomsk oblast will be served by the two largest tour-of-duty settlements of Novyi Vasyugan and Pionernyi, which will be staffed by workers with permanent houses in Strezhevoe. Pionernyi is currently the largest TOD settlement, and will eventually become a city. All the flats and buildings are joined by covered pavements enabling the Russians to describe it as a city under one roof. Unlike the conventional TOD settlement, it is expected to last for decades. The first stage of its construction has been completed.

Novyi Vasyugan, 200 km south-west of Strezhevoe, will serve the Vasyugansk field which is expected to be producing up to 6 mn tons/year shortly. If a permanent settlement had been built for 20,000 workers, it would have cost 500 mn roubles, and the oilfield would not have begun operating until 1983. The TOD settlement cost only 150 mn roubles and, after the decision was taken in 1977 to develop the field, the oil began flowing in 1978. Like Pionernyi, all the pavements will be covered, and the settlement will be semi-permanent.

TOD settlements are also being built in the gasfields. That at Gaz-Sale was designed by Leningrad architects, with the flats built up top of the shops, so that the inhabitants do not need to go outside in winter. It seems similar in structure to those at Pionernyi and Novyi Vasyugan.

Sociologists have been researching into rates of labour turnover among TOD workers. In none of the applications for transfer were bad living conditions cited as a reason. The main problem arose with workers who worked for five or six days with two days off; the rate of labour turnover fell by 67 per cent for workers working for 15 days with five days off. Accordingly, the workers at Novyi Vasyugan now work for up to 15 days before returning home to Strezhevoe.

Most workers wanting transfers were those without permanent flats in Strezhevoe. As the volume of oil from the Vasyugansk field will eventually exceed that from the Strezhevoe area (i.e. Sovetsk field), far more flats will need to be built in Strezhevoe for TOD workers. In 1979, the town got 40,000 sq m of new housing, and the plan for 1980 was set at 50,000 sq m.

As well as housing, the provision of amenities is crucial to attracting workers to Tyumen oblast. In spite of repeated complaints about the slow rate of construction, Tyumen cities seem to be adequately supplied, and the plan for 1981 calls for schools with 30,000 places, pre-school institutions for 24,000 children, and hospitals with 3,550 beds to be built.

The cities of the gasfields pose special problems due to their inaccessability, and this problem will not be remedied until the Surgut-Urengoi railway has been extended to Novyi Urengoi and Nadym.

Novyi Urengoi is probably the fastest growing city of Tyumen oblast, and though there has been an administrative shake-up resulting in the creation of a separate organisation charged with building housing, the builders are hindered by the failure of the LenZNIIEP Institute of Leningrad to complete the city's General Plan. The town now covers a large area - 14 kms by 4 - but not one of its microrayons has yet been consolidated.

However, the situation is improving. After 7,000 sq m of housing were built in 1979 and 70,000 in 1980, the 1981 Annual Plan was set at 180,000 sq m, and subsequently raised to 220,000 in July 1981. The actual outcome was 205,000 sq m (sufficient for 16,500 people), including a large number of 9-storey blocks built by workers of the Lenurengoistroi trest of Leningrad. This trest has promised to build 180,000 sq m during the 1981-1983 control period, accounting for 20 per cent of all new housing in Novyi Urengoi.

At the moment, Novyi Urengoi consists basically of five-storey flats built in the form of polygons with all the schools, kindergartens, shops, cultural and medical facilities built in the middle of the polygons. It has a hospital, several polyclinics, a shopping centre, a bakery, a 25-book library and an airport. Immediate plans call for the construction of more bakeries, a creamery, a multiple store and a vegetable store. The city is endeavouring to reduce its food imports by building several battery farms for pigs and hens, and a 4,000 sq m hothouse heated by waste heat from a gas preparation plant.

The creation of Nadym began as recently as 1972 to assist the development of the Medvezhe gasfield. Its population quickly grew to 15,000 housed in five-storey flats. It has an airport, a cinema, two schools with 1,176 places each, and several kindergartens and playschools. These are necessary because of the age structure of the population; the average age of Nadym workers is 27, and a third of the population are children. Nadym also has a laundry, a drycleaning shop, public baths, several shops and restaurants, a central boiler-house providing hot water for central heating in flats, and a sewage disposal plant. However, there is said to remain a disproportion between industrial infrastructural construction, and people soon leave because many necessary amenities do not yet exist. There are several reasons for this failure. The need to develop the Medvezhe gasfield as quickly as possible meant that Tyumengazprom exhausted its resources on that project, with the result that two years after gas production reached its peak the city centre of Nadym was still an empty space. There were also problems with a shortage of qualified workers, the severe climate, and sporadic supplies of materials. A brick factory producing 20 mn bricks a year has been built, but bricks are still being imported at a cost of 1 rouble each from Novosibirsk.

The planners have disagreed on how far and how rapidly Nadym should be developed, and these problems have afflicted almost every other town in the Siberian oil and gas regions. This is especially apparent where the provision of kindergartens is concerned, a major task in Western Siberia because of the population's age structure.

In order to carry out the requirements of the 1980-1983 control period for the construction of housing and amenities, Minneftegazstroi has set up three new construction trusts. These are Urengoigazstroi (for the gas-producing region), Obneftegazstroi for the Middle Ob region, and Priuralneftegazstroi for the lower reaches of the Ob where oil and gasfields will be developed during the 1980s.

TRANSPORT

Prior to 1975, equipment for the oil and gasfields was delivered by river, either down the Ob from Novosibirsk, down the Irtysh from Omsk, or down the Tobol from Tyumen. Loads were also carried by rail from Ivdel to the railhead on the Ob at Sergino. However, the rivers were ice-free for only five summer months each year, and after being unloaded at the wharves of Surgut, Nizhnevartovsk, Nefteyugansk etc., freight had to wait for the winter months when it could be hauled by tracked vehicles to its destination. This was because during the summer the entire Western Siberian Plain becomes a swamp which remains impassable until it freezes over.

After several years' argument about whether or not the fuel potential of Western Siberia justified its construction, work began on a single-track railway line connecting the Trans-Siberian railway at Tyumen with Surgut. Even after its construction had begun, it was argued by many that the steamship organisations were capable of handling all the required freight, and the railway was dubbed "the line to nowhere". After passing through Tobolsk, the tracklayers reached Surgut in 1975, and this enabled stores of materials to be reduced from 300 days supply to 150 or less.

In the extremely short period, given the conditions, of 1½ years, the line was extended for 186 km to Nizhnevartovsk, and the decision was taken to build another line northwards to Urengoi. Construction of this line by the Tyumen-stroiput trust began in November 1976 at the station of Ortyagun on the Surgut-Nizhnevartovsk line, 57 km east of Surgut. Settlements have been built at intervals of roughly 100 km: Kogolymsk after 91, Noyabrskaya after 204, Khanomei after 280 and Purpe after 380 until the track reached the small settlement of Tarko-Sale (population 8,000) at 455 kms in April 1980. The plan called for the last 122 kms to Urengoi to be completed by the end of 1980, and this was achieved with the construction rate reaching 485 metres a day - an extraordinary rate given the need to cross 100 rivers.

The line is now being extended for 85 kms westwards to Novyi Urengoi, and this track is planned for completion in the second quarter of 1982. Novyi Urengoi is currently connected by rail with the river port of Nadym. This 200-km line was part of a grandiose scheme by Stalin to build a line across the north of the USSR. It was abandoned after his death, but the Ministry of Gas has taken it over as a local line. However, the Nadym port serves over 100 customers, and was able to handle only 750,000 tons of freight for Novyi Urengoi in 1980, although the railway line can handle up to 6 mn tons a year. This problem cannot be solved by expanding Nadym's port facilities, because it is the width and depth of the River Nadym and the short navigation season (120 days from June to September) which limit the volume of freight that can be transported through Nadym. It makes sense to extend the railway to Salekhard, thereby enabling Novyi Urengoi to be approached from two directions, but this project is still at the discussion stage.

The Ministry of Transport Construction (Mintransstroi) has to decide how to spend its rather slender resources in northern Tyumen. While a Salekhard-Nadym railway will assist the development of Urengoi gasfield, it is equally important for Yamburg to be connected up to the rail network, and another possibility is that the track may point in a north-easterly direction towards the Russkoe deposit, and eventually reach the River Yenisei at Igarka before continuing northwards to Dudinka. This would effectively link Norilsk with the rest of the USSR.

The Tyumen-Surgut-Nizhnevartovsk railway carried 10 mn tons of freight in 1980, or 100 per cent more than in 1978, and it is to be improved so that in 1985 it can handle 20 mn tons a year. While the line may be double-tracked eventually, it appears that current plans expect the congestion problem to be overcome by the construction of 200 kms of branch lines and a large number of sidings, especially on the Tobolsk-Surgut bottleneck.

The organisation of rail traffic will be improved. Delays in loading and unloading are reported to have tied up rolling stock to the extent that "time and again the railway has been literally paralysed", according to Gosplan's deputy chairman, and railway wagons are idle for 13 hours a day instead of the normal eight. During 1978, loading bays at Tyumen, Tobolsk and Omsk held 170,000 tons of freight awaiting collection because of a shortage of wagons. Consequently oil workers were receiving new rigs up to a year after they were promised.

The volume of freight carried by river amounted to 19 mn tons in 1981, or
12 per cent more than in 1980, and is expected to rise to 23 mn tons in 1985.
The steamships will retain their leading role in supplying the oil and gas
producing regions. Most of this traffic sails down the Irtysh, with Omsk
river workers moving 8 mn tons of freight destined for the oilfields in 1980,
or 1 mn more than in 1979. Some 2 mn tons was carried down the River Nadym to
Nadym City in 1981 compared with 1 mn tons in 1980, and 0.5 mn tons went down
the River Pur to Urengoi. Traffic on the Pur proceeds with difficulty, and
as the volume of freight is planned to rise to 5 mn tons in 1985, the Irtysh
Steamship Line has begun a project to make the Pur navigable as far south
as Urengoi by dredging it and removing sandbanks. This task seems to have
been insuperable, because the emphasis is now on producing ships with shallower
draughts. The Novosibirsk shipbuilding plant has begun the construction of
a new type of motor tug-boat which can tow barges in half a metre of water.
During the Eleventh FYP, new ports will be built at Nadym, Urengoi, Sergino and
Nizhnevartovsk.

The problem of seasonality remains. While Sanders of Bechtel has said that
"the amount of traffic on these rivers in summer is astounding"[1] they remain iced
up from mid September to mid April, and normal navigation does not begin on
the Arctic sea route until June. The Arktika and Kiev atomic icebreakers clear
paths for vital loads for the gasfields in winter, and year-round sea traffic
has been possible for two years now. Some 2.5 mn tons of freight is to be
delivered to Yamal customers by sea in 1980. Icebreakers are also being used
to clear paths down the rivers.

Most road traffic in the oil and gas regions travels in winter along winter
roads made from compacted snow or ice held together with logs. They are built
every December to otherwise inaccessible settlements like Novoagansk, Povkhov-
skoe, Varyegan and Nonyegan. An ice road is even made across the frozen Ob
Gulf from the Yamal settlement of Se-Yakha to the Gyda Peninsula so that
geologists can be supplied from the Yamal base of Kharasavei. The standard
vehicle for the winter road is the Buran snocat, made by the Tybinsk (Yaroslavi
oblast) engineering plant, which can work in temperatures of -50 degrees. It
has a 35 HP engine, and can carry 15-ton loads.

The construction of permanent all-season roads is proceeding slowly. Some
2,146 kms of roads were built between 1970 and 1979, including 261 km in 1979,
but the Eleventh FYP target calls for 4,000 to 4,500 kms to be built over 1981-
1985, including 2,000 kms with hard surfaces. The plan includes 575 kms in
1981 (about 400 kms were built), 710 in 1982 and the same number in 1983. This
leaves at least 1,000 to be built in 1984 and 1,500 in 1985. By 1985, the
Khanti-Mansiiskii and Yamal-Nenetskii regions are planned to have 15,000 kms of
roads, including 4,200 with hard surfaces.

Tyumen's roads are built by the Tyumenavtodor directorate, belonging to the
Ministry of Transport Construction. In recent years, it has spent 1,184 mn
roubles on building roads as well as setting up 16 new construction and repair
bases and a bridge-building trust. It meets its plan targets every year,
but the plans are said to be inadequate. Of all the new oil deposits developed
since 1975, only ten were expected to have roads by 1980, and of the ten
deposits which began operating in 1976-1977, not one had a road by 1979.

---

1 Oil and Gas Journal 21/4/80

The situation is even worse in the gasfields. The Medvezhe deposit has been working at full capacity for two years now, but it still has only half its projected length of hard surface roads. At Urengoi, 45 km of roads should have been built so far, but it only has 5 km, and the plan for 1980 was set at 20 km of new roads when a further 50 are needed. The installation of gas-processing units, without which gas cannot be produced, is being delayed because of the shortage of roads. Urengoi's roads are particularly important because they will provide an approach to the new Yamburg deposit, and the situation has become so desperate that workers from Uzbekistan have been drafted in to build 60 kms a year.

Oil workers have complained about having to work in mud-baths, and then finally abandon the exhausted deposit along a newly built road. In 1979, they asked the government of a further 400 km of roads to be built above plan, and the government responded by drafting in thousands more Komsomol volunteers to build roads. However the quality of work has been declining, with a third of all roads needing repair within five years.

As well as serving oil and gas deposits, roads must be built alongside pipelines to facilitate repairs. In summer, most pipelines can be reached only by helicopter, and this means that breakages take up to ten days to repair. During this time, large volumes of oil or gas can be lost, and the cost of building pipeline roads can soon be recouped.

The ambitious roadbuilding plans for 1981-1985 can only be achieved if supplies of materials are improved. While Tyumenavtodor has built its own bitumen plant of capacity 100,000 tons/year, it also depends on the Ministry of Construction Materials sending 1 mn tons of road metal and waste each year from the Uralsasbest asbestos plant in Sverdlovsk. However, over four years, the deliveries from this plant were underfulfilled by 2.5 mn tons and the Irbit and Nazyvaevsk brick plants have hopelessly underfulfilled their targets for the supply of bricks.

The latest experiments in the use of new materials involve the use of wastes from synthetic fibre plants for building roads across swamps. In 1976, an experiment was carried out using waste instead of logs covered with earth, and although it was highly successful, costing 10 per cent less, the idea has not gone beyond the experimental stage due to a shortage of material.

The most common truck used in Tyumen oblast is the Czech-made Tatra, known by the Russians as the Severyanka. Some 30,000 are now operating, including 12,694 imported during the last FYP. During the Eleventh FYP imports will rise to 4,000 a year. The USSR's first plant for repairing Tatras has been set up at Nizhnevartovsk. It is expected to carry out 600 full overhauls a year plus repairs to engines. Although construction of the plant began in 1977, it was only completed in 1981 due to shortcomings by the Soviet builders of Megiongaz-stroi.

Soviet built trucks tend to be more specialised, such as the Belaz 549S 75-ton tipper truck. It has frost-resistant tyres and a protected engine, and can operate in -60 degrees. Production is taking place at the Minsk lorry plant, which also produces 120-ton lorries. The Ulyanovsk lorry plant produces a cross-country vehicle able to carry loads of 0.8 tons, and the Volokolamsk bus factory makes a specially designed bus for northern conditions.

Fuel for road transport has to be brought in from the Omsk refinery. This pro-duces acute difficulties in the gas-producing region, where 1.5 mn tons of fuel and lubricants were needed in 1979, and 8 mn roubles were spent on transporting them. But Urengoi gas deposit already has an experimental plant for refining gas condensate, and in the future it will produce sufficient petrol, diesel and lubricants for all the road traffic on the gasfields while the surplus condensate will be sent to the Tobolsk petrochemical plant by pipeline. During tests of the plant, it produced 1,000 tons of motor fuel, providing savings of more than 1 mn roubles. It is also planned to build a number of small mobile condensate refineries able to produce diesel and lubes.

Most settlements in the oil and gas regions have an airport able to receive MI-8 helicopters and AN-26 and AN-12 cargo planes. Some can also take the giant IL-76 transport planes. Most passenger traffic is by plane because of the immense distances and the Tyumen civil Aviation Directorate carried 4.3 mn passengers in 1976.

POWER

So far, the Tyumen oil and gas region has only one large-scale source of power, the Surgut power station of capacity 2,730 mw. The station has 13 units, each of 210 mw, and its size has been built up gradually since the first unit gave power in 1969. Six units were operating by the end of the Ninth FYP period, and a further six were operating by the end of the Tenth. The 13th was commissioned in 1981, and two more are to be installed, bringing the station to its rated capacity of 3,150 mw. It runs on dry casinghead gas from local gas refineries, and gas from the gas cap at Fedorovsk oil deposit. Work has already started on the foundations of a second Surgut power station. Its capacity will be 4,800 mw, made up of six units of 800 mw each, and it will start to give power in 1983.

While the construction of the power stations has gone according to schedule, work on transmission lines and sub-stations has lagged behind. The most important transmission line is that of 500 kv to Megion, where the power is distributed to users in the Nizhnevartovsk region like the Samotlor oil deposit. This line was completed only in 1977, by which time sub-stations of 35, 110 and 220 kv capacity had been built, although the first 500 kv sub-station was still under construction.

Another 500 kv line has been erected by Zapsibelektstroi to connect Urengoi gasfield with the Surgut power station, but delays by the survey organisation, the Sverdlovsk branch of Energosibproyekt lead to work falling behind schedule. The Surgut-Kholmogorsk section was completed on time, but survey work for the Kholmogorsk-Tarko Sale section was finished only in July 1979. By April 1980, the line was delivering power to the Muravlenkovsk oil deposit, and it reached Urengoi in early 1982. Surgut power station also delivers power to the national grid through two 500 kv lines which run via Demyansk to Tyumen.

Since 1965, 10,000 km of high voltage transmission lines have been erected in Tyumen oblast, but in the oil and gas regions the work rate has accelerated only slightly, from 547 km in 1978 to 571 in 1979. Consequently, many oil deposits came on-stream long before they were served with electricity. The situation was good until 1975, when 2.62 mn metres of drilling from a total of 2.79 mn was accomplished with the aid of electricity (the rest was performed with diesel drive). But of the ten deposits introduced during 1976-1977, only two had a permanent electricity supply by the end of 1978. With the exception of two distant deposits, they were all supposed to be connected with the grid within a few months of coming on-stream, but had to make do with mobile diesel stations instead. This held back their development to the extent that the ten deposits produced only 2.5 mn tons of oil in two years.

The Severo-Varyegan deposit, due to become the second largest producer after
Samotlor, was supplied with electricity only in July 1977. The urgency of the
task was so great that helicopters were used to transport the pylons. The
power shortages afflicting the Nizhnevartovsk field (of which Varyegansk is part)
have been eased since April 1981, when a 220 kv line running 900 kms from the
Kuzbas through Tomsk reached Strezhevoe, and was connected to the local power
distribution system. This had previously been served only by the Surgut
station, and the power supply was unreliable due to the overloading of the
Surgut-Megion line.

The problems are even greater in the gasfields. The Ministry of Power was
supposed to erect 273 km of 110 kv line in 1977, but only 46 km were provided.
In the absence of the link with the Surgut power station, the northern gasfields
were powered by a multitude of small mobile stations. More than 700 workers
were employed on these small stations at Urengoi alone and a mobile station had
to be commissioned for each new gas treatment plant, compressor station, drilling
rig, residential area etc., and it has been estimated that lower drilling
rates due to the use of diesel rather than electric rigs were leading to a failure
to drill 50 wells a year. Now that the Surgut-Urengoi transmission line has been
completed, it is expected that many of these small stations will be closed.

A large-scale gas-fired power station is to be built during the next few years
to serve the Urengoi gas field. With a rated capacity of 4,000 mw, it is to be
located at the new town of Tikhii, and since June 1981, local officials have
been urging that its construction be started immediately.

Floating power stations are a good way of providing power until Urengoi power
station has been built, although the Severnoe Siyanie-5 has been at anchor at
Nadym for two years now without producing electricity. There is nothing wrong
with the ship; it is an improvement on its predecessors, which are working
satisfactorily. It was ordered in 1974 and arrived in 1978, but only in the
summer of 1980 were the necessary transmission lines and substations installed.

The provision of centralised thermal energy sources is also lagging behind.
Even Surgut, with over 150,000 people, does not have a centralised supply, but
is served by 25 small boilers. Glavtyumenneftegaz administers over 900 boilers
with a total capacity of 1,465 G-cals an hour. They use 600,000 tons of standard
fuel a year, including 260 mn cu m of gas and 169,000 tons of fuel oil. All new
boilers are designed to use only gas, and large new boilers with capacities
of up to 100 G-cals per hour are being built, each one of which will enable
60 small inefficient boilers to be scrapped.

CHAPTER 9

The Caspian Sea Region

DEEP WATER TECHNOLOGY

During the Tenth FYP period, the Russians have been modernising and augmenting their offshore drilling capacity, although not as rapidly as planned. They currently have seven jackups (four modern and three obsolete) and two semi-submersibles. Slow progress in developing their deep-water capacity prompted a radical reorganisation in 1978, when all offshore work was given to the Ministry of Gas. It immediately set up a new organisation, Kaspmorneftegazprom, designed to concentrate on working the 22 oil and gas deposits in the Caspian, although it has also been exploring in the Black Sea.

The four modern jack-ups are the "Bakii," the "60 let Oktyabrya", the "Sivash" and the "60 let Azerbaidzhan." The latter began operating in 1981, and was sometimes referred to before it was launched as the "Bakiyets". The obsolete jack-ups are the "Azerbaidzhan", the "Apsheron" and the "Khazar".

The Bakii appeared in 1975, and like all the modern jack-ups, was built by the Krasnyi Barrikady shipyard in Astrakhan. It was sent to the Yuzhnaya-2 deposit, where it was set to work on a 5,300 metre well. By Septembre 1978 it had drilled to 3,300 metres at an apparent rate of 750 metres a month, which is fast by Soviet offshore standards. It has been described as "quite efficient" by a Western expert,[1] and can drill to 5,000 metres in 75 metres of water.

The "60 let Oktyabryua" was introduced to service in November 1977, and is designed to drill to 6,000 metres in 75 metres of water. With legs of 108 metres it is 8 metres higher than its predecessor the "Bakii". It was built in Astrakhan with equipment from Uralmash and was towed out in December 1977 to the Banka Andreyeva deposit, where it began drilling a 6,000 metre well. By May 1978, it had reached 5,240 metres at the very impressive rate of over 1,000 metres a month.

In 1979, the third modern jack-up, the "Sivash" began operating in the Black Sea, looking for gas in the Ilichevsk deposit off the Crimea. It is of the same "Kaspyi" class as the other two, and can drill to 6,000 metres in 70 metres of water. The "60 let Azerbaidzhan" was launched in April 1981 at Astrakhan and sent to join "60 let Oktyabrya" at the Banka Andreyeva.

According to the Tenth FYP, ten modern jack-ups were to be completed over 1976-1980, but only four were completed. With the Soviets preferring to concentrate on semi-submersibles that can drill in much deeper water, tne eleventh FYP has set a target of five jack-ups to be built over 1981-1985.

The first Soviet semi-submersible was the "Kaspmorneft". It was assembled in Astrakhan from equipment produced by Armco and the Rauma Repola shipyard of Finland. It can drill to 6,000 metres in 200 metres of water. It weighs 17,000 tons, and has two 79 metre pontoons. Assembly began in December 1977

---

1 World Oil 15/ 8 /79

and was completed in November 1979. It has an underwater television system,
pollution-prevention equipment, a helicopter pad, accommodation for the crew, mud
pumps and well-cementing equipment, as well as the drilling rig. It was towed
to the giant "28th April" deposit in early 1980, where it is moored by eight
anchors of 18 tons each. It is now drilling a 4,500 metre well in 140 metres
of water, a record for the USSR.

The construction of the first semi-submersible built entirely by the USSR
began at Astrakhan in December 1978. Known as "Shelf-1" it was completed in
October 1980. It was towed to Baku in July 1981, and after final fitting out
was sent to a new exploration area believed to be an easterly extension of the
"28th April" deposit, where it is drilling to 4,500 metres in 200 metres of water.
Its sister vessel, "Shelf-2" was launched at the Moskii shipyard in Astrakhan in
July 1981 and is now being fitted out in Baku. The Shelf rigs are owned by
Kaspburneftegazprom, and are entirely domestically produced with equipment from
Uralmash of Sverdlovsk, Barrikady of Volgograd and many other suppliers. They
have the same characteristics as the Kaspmorneft, and can withstand Force 7
storms. They are the first in series for which the Astrakhan shipyards have
been reconstructed, and two more are to be built in the next few years. By
1985, it is planned that 13 floating rigs will be operating in the Caspian,
allowing the annual volume of exploration and appraisal drilling to rise from
350,000 to one million metres during 1981-1985.

Most Soviet platforms are built by the "Oktyabrskaya Revolyutsiya" engineering
plant of Baku and are a standard 57 metres high, although the plant has made
some bigger models. The biggest platforms built at the present time weigh.
2,100 tons and are designed to stand in 100 metres of water. They are built in
two halves, each of which can carry a rig able to drill 12 wells. In
December 1981, the first such platform was completed at the "28th April"
deposit.

Meanwhile, a new platform assembly plant is under construction at Karadag on
the Caspian coast south of Baku. When it is finished in 1983, it will produce
three platforms a year, weighing 20,000 tons each. They will stand 100 metres
high, and can be jacked up to stand in 200 metres of water. Eventually, platforms
from which 24 wells can be drilled in 350 metres of water will be built at
Karadag. The plant will cover 200 hectares - 120 on land and 80 at sea - and
construction is proceeding ahead of schedule. It is being built by Entrepose
of France at a cost of Fr 500 mn.

Use is beginning of dynamically self-positioning exploration ships. The first
of these vessels, Spetssudno-1, was built by the Dubigeon-Normandie Yard,
Rouen. It has a system of propellors which are switched on and off automatically
depending on the direction of the waves, the current and the wind, and keep the
ship permanently in the required position. It employs a revolving core-sampler,
designed by the Leningrad Mining Institute, suspended from the ship on a cable
and powered by an electric motor. It can drill holes up to 200 metres deep
in up to 600 metres of water; it photographs the sea-bed and looks for places
where jack-ups can be erected. It began work at the "28th April" deposit in
June 1980.

The new geophysical exploration ship, the "Suleiman Bagirov" has recently been
working at Neftyanye Kamny. Where as previous vessels could only survey to
depths of 4,000 metres below the sea-bed, the new ship can carry out surveying
in combination with other geophysical methods to depths of 10,000 metres.

Wellhead valves for marine wells are made at several plants, notably the
"Krasnyi Molot" engineering plant of Groznyi, and the Industrial Rubber Plant
at Volzhskii makes items designed to insulate well-head equipment from water.

Experience gained in the construction and operation of marine rigs in the Caspian and Black Sea will eventually be utilised in the much more difficult conditions of the Arctic Sea. The construction of the first Arctic semi-submersible "Vyborg-1" has already begun at the Vyborg Shipbuilding Plant north-west of Leningrad. Weighing 20,000 tons, and able to drill to 6,000 metres in 200 metres of water, it is basically of the Shelf class being built at Astrakhan, but it will be very much stronger so as to withstand Arctic conditions.

GASLIFT

Gaslift is being used more comprehensively in the Caspian fields than anywhere else in the USSR because of the substantial yields of gas from many of these fields, and because waterflooding is proving ineffective in the heavy oil deposits.

The successful use of gaslift at the Azerbaidzhan offshore field of Neftyaniye Kamny has enabled exhausted wells to be reactivated at a rate of 10 tons a day, and at the Sangachaly-More group of deposits, 16 gas distribution batteries have been installed. They are built on platforms, and can pump gas back into the oil-bearing strata at a pressure of 100 ats.

Gaslift is much more extensive in the onshore Turkmen fields. In the past, waterflooding has been the main secondary recovery method and is still proving very effective at some sections of Kotur Tepe. But by the beginning of 1978, 400 wells had been converted to gaslift, which is said to be four to five times as effective as pumping. The plan was to convert 1,000 wells by the end of 1980, using the closed cycle method so as to preserve the gas, but this plan was considerably underfulfilled, thereby leading to a decline in oil production.

The rapid fall in pressure at some gaslift wells had led to their underproducing by up to 10 tons a day and requiring 30 per cent more gas. This problem has been alleviated by the invention of a new type of valve, which automatically regulates the pressure and the volume of gas.

Scientists from the Nebit Dag Institute have elaborated a method of using pools of natural gas with pressures of 160 ats to raise oil from neighbouring pools with very low formation pressures. This permits the construction of compressor stations to be avoided, and it was planned to extract oil from 50 wells by this method of by 1980.

Most Soviet research into the effectiveness of gaslift of heavy oil conditions has been carried out on the Mangyshlak Peninsula, where 700 wells had been converted to gaslift by the end of 1977, mainly at the Uzen deposit. At present, Uzen is worked with gaslift and waterflooding; 37 per cent of the wells work on gaslift and 53 per cent are mechanically pumped.

The gaslift system is based on a compressor station of capacity 1.7 bn cu m of gas a year, with a working pressure of 5.3 to 5.4 mpa at the Kazakh Gas Refinery. The 594 gaslift wells had an average yield of 70 tons a day with a water-cut of 25 per cent and accounted for 60 per cent of all Uzen oil in 1975.

Since its gaslift system was introduced in 1969 (and completed in 1975), Uzen has served as a testing centre for Soviet gaslift equipment. New types of valve and modular automatic gas distribution batteries have been tested here, and the system is said to have worked well. The problems have mainly been breakdowns by the compressor station and the gas pipelines, due to an excessive moisture content of 0.6 grams/cu m in the gas from Tengy deposit and 3 grams/cu m in that from the Kazakh refinery. The pipelines have consequently suffered from hydrate formation and condensate blockages.

Another problem has been the sharp rise in the hydrogen sulphide content of casinghead gas, due to the use of sea-water containing bacteria. The gaslift scheme was designed on the assumption that the gas would not contain hydrogen sulphide, and it has proved necessary to close down some gas collection plants and pump bactericides into the deposit. A sulphur separation facility is being built at the refinery.

As well as the technical problems, there are doubts about the cost effectiveness of gaslift at Uzen. It has been shown that the cost of gaslift is lower than that of pumping only for wells yielding more than 100 tons of fluid a day with a water-cut of less than 30 per cent. In other cases, pumping is cheaper, and for wells yielding only 5 tons a day with a watercut of 50 per cent, it is substantially cheaper because gaslift requires a large volume of gas to raise a small volume of liquid. For a well-yield of 5 tons a day, 1,100 cu m of gas are required per ton of fluid, but for a yield of 50 tons/day only 115 cu m are needed.

On the basis of these results, 80 gaslift wells at Uzen with low yields and with an average gas requirement of 300 cu m/ton were transferred to mechanical pumping. On the other hand, 30 free-flowing or pumped wells with high yields were transferred to gaslift. Thanks to these measures, 104,000 tons of extra oil were obtained in 1977. In 1979, Uzen was said to have 125 wells which should be transferred from gaslift to pumping.

Electric pumps cannot be used because of paraffin deposition, salt accumulation and large volumes of gas. Therefore sucker-rod pumps have to be used.

STEAM INJECTION

Experiments in steam injection have been carried out at Kenkiyak deposit in the Emba field since 1972. The deposit came on-stream in 1969 with mid-Jurassic oil having a high gravity of 0.915 grams/cu m and a viscosity of 180 to 250 centipoise. Average well yields soon fell from 13 to 4.5 t/d.

In 1972, three steam injection wells were drilled in a section of the deposit with an average well-yield of 6.3 t/d. By 1979, yields averaged 6.1 t/d, but it has been calculated that without steam injection the yield would have fallen to 2.5 t/d as at other sections. Some wells gave 22 t/d, which is high for such heavy oil, and could have given more but for sand invasion which necessitated lengthy shutdowns for cleaning.

The steam is injected at a temperature of 250 degrees and at a rate of 73 tons per day per well. During 1978, 127,000 tons of steam were injected and 70,000 tons of oil obtained, 48,000 more than without the use of steam. While a significant amount of extra oil has been obtained by this method, there is no indication of its cost effectiveness. The new reference prices for oil to be introduced during 1981-1982 should encourage production organisation to make more extensive use of this method, especially when the serial production of a new type of steam injection unit, developed at Tuimazy, begins. A factory for this purpose is now under construction in the USSR.

MINING

It is intended to mine highly viscous oil at shallow depths in up to 25 deposits
in Azerbaidzhan.   Preparations have begun at the Leninskii deposit at Balakhany
which has been worked for over a century.   Two shafts are to be sunk to a depth
of 340 metres by the Shakhtospetsstroi organisation from the Donbas coalfield.
The oil will flow into chambers in the shafts, from which it will be taken to
the surface.

CHAPTER 10

Production Forecasts to 1985 and 1990

The Eleventh FYP targets for oil and gas production in 1985 of 620 to 645 mn tons of oil and 600 to 640 bn cu m of gas were accepted by the 26th Party Congress in February 1981, and confirmed by the Party Plenum of November 1981 at 630 mn tons of oil and 630 bn cu m of gas. These targets, particularly that for oil, were considerably higher than most Western analysts had anticipated because it was widely believed that Soviet technology was too backward to permit the Russians to overcome the formidable problems involved in raising oil output beyond the level of 600 mn tons a year.

The EIU report "Soviet Oil and Gas to 1990" was one of the very few Western reports to believe that the Russians had the technology (or were able to create it) to raise hydrocarbon output so as to fulfill their domestic needs, cover most of Eastern Europe's requirements, and remain an important exporter to the West. In forecasting a 1985 target of 700 mn tons of oil and 600 bn cu m of gas, totalling 1,708 mn tons of standard fuel, the EIU forecast came to within 1.8 per cent of the 1,678 mn tsf expected by the higher range of the official five year plan, and only 3.9 per cent higher than the concretised target of 1,644 mn tsf.

In fact, the forecasts of the EIU report accurately reflected official Soviet thinking as late as summer 1980, when Gosplan was drawing up variants of possible fuel output in 1985, and relating them to planned economic and technological developments in the rest of the economy, and anticipated trends in the world economy.

In particular, it was not clear how quickly gas could be substituted for fuel oil (the constraint being the rate at which distribution pipelines could be laid rather than the cost or duration of the process of converting an oil-fired facility), how quickly new secondary refining capacity could be installed, how reliable foreign sources of wide diameter pipe for gas pipelines may prove to be, and whether the Yamal pipeline deal would come off or whether the Soviets would have to try to sell more oil. Could the Western economies, in fact, absorb more oil in view of their economic recession which was beginning to occur in the summer of 1980.

Gosplan's final decision may well have been influenced by the likely conclusion to negotiations with Western banks for credits for the proposed Yamal gas pipeline. At that time, there was every chance that a West German banking consortium would provide a 20 bn DM loan for the pipeline, and this would enable the Soviets to eventually pay off their debts to the West with gas rather than oil. In Western Siberia, it is much cheaper to extract a given volume of gas than the same volume (in terms of comparable calorific value) of oil; the drawback to gas production is the extremely high cost of providing the means of transportation. With the West Germans apparently prepared to lend the cost of the Yamal pipeline, the gas option became much more attractive to the Russians.

It is apparent that during the autumn of 1980, Gosplan made their minds up to devote more resources to gas and fewer to oil than was originally planned. An indication of this is the subsequent revision of the plan to drill 20 mn metres of development wells in Western Siberia in 1985 to only 18 mn metres. They have also made significant changes in their Siberian oilfield development strategy. In order to reach an original target for Western Siberia of 450 mn tons a year, the Russians intended to carry out the intensive working (i.e. with dense well networks) of as many as 40 new oil deposits. Some unofficial targets floated during 1980 for the construction of roads, oilfield pipelines, electricity transmission lines and tour-of-duty settlements were met with disbelief by the construction organisations which would be charged with trying to fulfill them if they were officially adopted. It was pointed out that, in spite of the more formidable problems facing the construction enterprises in the gasfields (such as the lack of an all-season transport link with the outside world), the gas industry could easily provide the entire increase in hydrocarbon output required over the 11th FYP period from just one deposit, Urengoi, compared with 40 small oil deposits scattered over a huge area of Western Siberia.

It was also felt that the reliability of supplies of large diameter pipe from Japan and West Germany was sufficiently high to allow gas production to go as high as 640 bn cu m in 1985, which would require 50,000 kms of pipe to be laid over 1981-1985. Accordingly, the planned number of new oil deposits to be exploited over 1981-1985 was scaled down to 27. Although these tend to be closer to the developed oilfields round Nizhnevartovsk and Surgut, thereby sharply reducing the volume of construction work in building up the necessary infrastructure, they are also much deeper with many of them giving oil from deeper than 4,000 metres. The development of comparatively shallow oil deposits in the Tarko Sale area, which will necessitate a huge increase in infrastructure costs has been postponed to the Twelvth FYP period 1986- 1990.

The gas option was unsuccessfully resisted by the Ministry of Gas. They asserted that Tyumen could produce no more than 320 bn cu m at the most in 1985 with the USSR yielding a maximum of 600 bn cu m. The adoption of the new strategy has meant that from a national total of 630 bn cu m anticipated in 1985, Tyumen oblast must produce 357 bn cu m, with all the increase over the original target coming from Urengoi.

The EIU report's forecast of 700 mn tons of oil in 1985 consisted of 460 mn tons from Western Siberia (441 mn tons from Tyumen and 20 mn from Tomsk oblast), 57 mn tons from the oilfields in or around the Caspian Sea, and 183 mn tons from elsewhere, mainly the Ural-Volga region. The Eleventh FYP target of 630 mn tons consists of 395 mn tons from Western Siberia, 52 from the Caspian fields and 183 mn tons from elsewhere. It is obvious that the Russians have scaled down their earlier expectations largely at the expense of the Western Siberian fields, because it is from here that the extra resources needed to push gas production to 630 bn cu m a year will be diverted.

It is useful to review the analysis behind the forecasts in the EIU report because it is still possible that the Russians may revert to their original intention to step up the exploitation of the Tyumen oilfields. The collapse of the Yamal deal is now most unlikely with the Russians involved in a war of prestige with the USA over its completion, and in any case the construction of the pipeline has begun with the first kilometre of pipe trenched on 12th February 1982. Yet, any one of a number of issues could provoke a new Western oil crisis, and if the spot price of crude were to approach (say) $50 a barrel with no prospect of it falling, then the Russians may be tempted to direct more resources into Western Siberia to take advantage of these prices. The analysis in the EIU report concluded that they could afford the cost in terms of human and capital resources to bring Siberian oil output to 460 mn tons a year in 1985, and 500 mn tons in 1990, and it is significant that Soviet planners and geologists are already talking about Tyumen oblast alone giving 500 mn tons in 1990.

All the projected increase in national output during the next decade will come from Western Siberia. The opening up of this region was once described by a Western observer as a "desperate gamble", yet all the evidence shows that the Russians do not take desperate gambles; that, if anything, they can be criticised for being ultra-careful and over-conservative. The delays in developing Western Siberia, caused by years of endless disputes over transport systems (rivers versus railways), habitation strategies (permanent settlements versus tour-of-duty settlements) the location of the engineering base (Urals versus Kuzbass) and many other issues provide convincing evidence that they waste valuable time by exhaustively studying every new question, and then try to make up for lost time by an all-out drive.

Table 30 shows that even with an increasing share of development drilling by Glavtyumenneftegaz devoted to injection wells, and with a big fall in the average yield of new oil wells, the USSR could expect to produce 440 mn tons in Tyumen oblast in 1985. It was assumed that the original plan for the drilling of 20 mn metres would be achieved, that 25 per cent of it would consist of injection well drilling and that the average depth would remain at 2,500 metres over 1979 to 1985 compared with 2,251 metres in 1975.

During the 1971-1975 period, the total output of one-year-old wells, which were entering the stock of old wells, was large enough to outweigh the fall in total output by those wells already in the "old well" category. This resulted in a rise in the average yield of all wells from 36,700 tons a year in 1971 to 45,500 in 1975.

During the period 1981-1985, a very large number of wells will be joining the stock of old wells, which would have grown from 6,591 in 1980 to 18,317 in 1985 under the assumptions of Table 30. However, the yields of these entrants would be very low; even as new wells, they would be producing only 77 tons a day in 1985 compared with 123 tons a day from new wells in 1980 and 168 in 1975. Consequently, their total output would be less than the fall in total output from existing old wells, and the average yield from all old wells would fall from 91 tons a day in 1979 to only 43 in 1985.

The rate of fall of output by old wells can be measured. In 1979, 1958 new wells produced 97.9 mn tons, but a year later these same wells might be expected to produce 78.3 mn tons. However, the total output by old wells rose by 14.8 mn tons during 1979-1980, suggesting that the existing stock of old wells lost 63.5 mn tons (some old wells, of course, were shut down). For 5,293 existing old wells to lose 63.5 mn tons in their annual output gives a rate of fall of output of 12,000 tons per well per year.

Table 30 suggests that the USSR may have expected 4,660 new wells to produce 146.3 mn tons in 1984, with output by old wells rising by 17.9 mn tons during 1984-1985. Thus it can be estimated that the rate of fall of output by old wells would amount to 6,500 tons per year. But while the rate of fall in 1979 represented 36 per cent of the average yield of an old well, by 1985 this coefficient would rise to 41.4 per cent. Therefore the rate of fall of output would grow significantly, and the proportion of a well's total output coming from its first few years of service would increase accordingly. In fact, most Siberian wells produce only tiny amounts of oil after their first five years of operation.

In 1975, the Ministry of Oil had 92,700 wells, of which 68,000 were oil wells, suggesting that injection and observation wells accounted for 26 per cent of the total. It is assumed that by 1985 the ratio of oilwells to injection wells in Western Siberia will be the same as for the USSR in 1975.

Table 30 also assumes that the rate of fall of output by old wells will increase. This phenomenon has been observed in the European part of the country, and Krylov attributes it to an over-dense well network (chapter 6). While development strategies for new deposits in Western Siberia envisage thinner well-networks, it has been assumed that this will prove insufficient to prevent a rise in the rate of fall of output by old wells.

Table 30 indicates how much oil the USSR could expect to produce if the original drilling target of 20 mn metres in 1985 was achieved and the original development strategy (involving 40 new deposits) was carried out. In fact, the EIU report asked if this drilling target was realistic, pointing out that it expected more wells to be completed during 1985 than during the first 3½ years of the Tenth FYP.

Under the new plan, the drilling target has been scaled down to 18 mn metres. The new development strategy involves a smaller number of wells being drilled to greater depths at more accessible deposits. This means that the average well depth will grow to 3,200 metres in 1985. Table 31 shows how the new plan is likely to raise the number of producing wells from 7,251 in 1979 to 20,125 in 1985 compared with almost 23,800 under the old plan. Consequently, production in 1985 is likely to amount to 380 mn tons rather than the 441 mn tons that could have been expected under the old plan.

Chapter 8 has dealt with the problems and prospects concerning labour, housing, transport and power in Western Siberia during the Eleventh FYP, and it is apparent that a big effort is under way to render the plan targets realisable. This includes bringing in a large number of drillers and other workers from other regions of the country.

The number of drilling crews in Western Siberia doubled during 1975-1980, and in 1979 alone it rose from 80 to 108. In fact it was supposed to rise by 47 with 22 new crews coming from the Bashkir, Tatar, Kuibyshev and Saratov regions (bringing their total number to 40 in 4 directorates), and 25 new crews being created by Glavtyumenneftegaz from local workers. During the Eleventh FYP period, five new directorates with 50 new crews are to be created in the Kuibyshev, Tatar, Bashkir, Ukraine and Belorussia regions. In addition, Glavtyumenneftegaz will set up a further four directorates, and perhaps another four later on. This suggests that at least 240 crews will be working in Western Siberia by 1985.

Crews from the Ural-Volga region began working in Tyumen in 1978, and by mid-1981 workers from the Tatar republic had drilled 1.2 mn metres in 370 wells. By the beginning of 1982, there were 18 crews of Tatar drillers working in Western Siberia. Bashkir drillers have been sent to Kholmogorsk, and Ukrainian workers from the Ivano-Frankovsk oilfield have been charged with helping to develop the Varyegan deposits. However, more outside drillers will be needed because the development to projected capacity of recently discovered giant deposits such as Severo-Pokursk, Urevsk, Pokacheva, Lokosovo, Kholmogorsk etc. will be accompanied by the development of 27 new deposits, thereby raising the number of deposits giving oil by 1985 to 67. The original variant foresaw the commissioning of 40 new deposits, including 15 by Nizhnevartovskneftegaz Association compared with 9 over 1976-1980, and the association was planned to drill 10 mn metres a year or half the total in the Tyumen oilfield. The Ministry of Geology will also need outside help in raising the volume of exploratory drilling from the inadequate levels of the past to 10 mn metres over 1981-1985 including 3 mn metres in 1985.

Table 30

Projected Output of Oil, Tyumen Oblast: original intention

	1979	1980	1981	1982	1983	1984	1985
Total development drilling (mn m)	6.1	9.0	10.1	11.7	13.9	16.6	20.0
Drilling of oil wells (mn m)	5.2	7.4	8.1	9.2	10.7	12.6	14.8
No. of new wells drilled	2,070	2,940	3,240	3,680	4,280	5,040	5,920
Stock of new wells, mid year average	1,958	2,505	3,090	3,460	3,980	4,660	5,480
Stock of old wells	5,293	6,591	8,251	10,299	12,592	15,229	18,317
Average new well-yield (tons)	50,000	45,000	41,000	38,000	34,800	31,400	28,000
Average old well-yield (tons)	33,300	29,000	25,000	22,500	20,000	17,700	15,700
Oil from new wells (mn tons)	97.9	112.7	126.7	131.5	138.5	146.3	153.4
Oil from old wells (mn tons)	176.3	191.1	206.3	231.7	251.8	269.7	287.6
Total oil production (mn tons)	274	304	333	363	390	416	441

Table 31

Projected Output of Oil, Tyumen Oblast: new intention

	1979	1980	1981	1982	1983	1984	1985
Total development drilling (mn m)	6.1	7.9	10.4	11.9	13.7	15.7	18.0
Drilling of oil wells (mn m)	5.2	6.5	8.3	9.4	10.5	11.9	13.3
No. of new wells drilled	2,070	2,500	3,090	3,340	3,640	3,970	4,160
Stock of new wells, mid yr ave.	1,958	2,285	2,795	3,215	3,490	3,800	4,065
Stock of old wells	5,293	6,591	8,051	9,838	11,840	13,905	16,060
Ave. new well-yield (tons)	50,000	47,300	43,500	39,000	35,000	32,000	29,500
Ave. old well-yield (tons)	33,300	29,500	25,000	22,100	19,750	17,700	16,200
Oil from new wells (mn tons)	97.9	108.1	121.6	125.4	122.2	121.6	119.9
Oil from old wells (mn tons)	176.3	194.4	201.3	217.4	233.8	246.1	260.2
Total oil production (mn tons)	274	302.5	323.0	342.8	356.0	367.7	380.1

The seasonality of drilling work creates special problems. According to Dunaev, the chief engineer of Glavtyumenneftegaz, the Russians like to build artificial islands, move equipment and assemble rigs in the winter months when the ground is hard, and concentrate on drilling during the summer.

It can be estimated that in 1979 the average volume of drilling per crew on the Tyumen oilfields was 65,000 metres. If this level of productivity is maintained, then 277 crews will be needed to meet the 1985 drilling target. It is reasonable to assume that higher productivity due to better equipment will outweigh the negative effects of more difficult conditions, and that productivity can rise by 3 per cent a year. In this case, 240 crews would be needed, and this is the number planned for 1985.

It is very difficult to provide forecasts of oil production from each deposit, as the Russians can choose from several variants of their current development strategy. They expect Samotlor to continue operating at its peak output level of 155 mn tons a year for the next five years at least, and some reports have even suggested that it will rise significantly. The Varyegan group is scheduled to produce 30 mn tons a year "some time in the future", and several other smaller deposits in the Nizhnevartovsk field will help to keep output rising beyond 1985. Kholmogorsk and its neighbouring deposits are planned to produce between 36 and 40 mn tons in 1985, with 25 mn tons coming from Kholmogorsk itself, and an additional pipeline is being built to the new Muravlenkovsk deposit. Severo-Pokursk, Lokosovo, Urevsk and Pokacheva are all believed to be giant deposits and therefore capable of each giving 30 mn tons a year in five years time, and the new Krasnoleninsk field, 200 kms west of Surgut, should be yielding useful amounts of oil by 1985. Under the old plan, output by the Surgut field was to double to over 100 mn tons in 1985, but this has probably been reduced to perhaps 70 mn tons while the Yugansk field should not produce much more than the 50 mn tons of 1981. Production by the small Shaim field is expected to double.

The Eleventh FYP expects production from the Caucasus, Kazakh and Central Asian fields in or around the Caspian Sea to grow from 46 mn tons in 1980 to 52 mn in 1985. The whole of this increase is anticipated from Kazakhstan, with the growth in output from the Azerbaidzhan side of the Caspian Sea covering a further fall by the Turkmen republic. It appears that the Russians expect their substantial investment in the Caspian Sea fields to be recouped in terms of gas rather than oil, with the yield by the newly discovered oil deposits rising only after 1985.

Output should also rise marginally in other small regions such as the Komi republic, Sakhalin Island, Georgia and Kaliningrad oblast. Elsewhere it will continue to decline, despite the much more intensive use of tertiary recovery methods, the drilling of wells down to 7,500 metres, and the exploitation of a large number of tiny deposits.

The volume of development drilling by the Ministry of Oil must grow from 15.8 mn metres in 1980 to 28 mn in 1985, or at a rate of 12.1 per cent a year. It is reasonable to assume that the annual growth in rig productivity of 3.8 per cent obtained over 1975-1980 will continue beyond 1985, and assuming no improvement in the usage rate of the ministry's rigs, their number will have to grow at a rate of 8 per cent a year.

Even if the anticipated doubling of the average bit run over 1980-1985 is substantially underfulfilled, the annual production of bits would not need to rise appreciably above the current level of 400,000. As in the past, the Russians will be improving the quality of their bits rather than trying to produce more of them. More down-hole motors and submersible pumps must be produced, as well as more (and better) drill pipe and casing pipe. Chapter 5 shows that they are undertaking these tasks.

Predicting forward to 1990 is more difficult. Soviet spokesmen have said that they do not expect Western Siberian production to stop climbing before 1995 or even later; Academician Trofimuk has claimed that annual output will be on the increase well into the 21st century. During the latter half of the 1980's, the so-called "second Tyumen" (the large deposits of oil lying at great depths beneath the northern gasfields) should start coming on-stream and off-shore drilling in the Kara and Barents Seas, which is due to start in 1983, should begin to give results. Development of the Eastern Siberian fields should not begin before 1990.

During the latter half of the 1980s, the main emphasis will be on natural gas and nuclear power for covering energy requirements, and the share of light and middle products in total product volume will be approaching the level currently attained in Western Europe. The production of oil will continue to rise, but only at marginal rates, and a figure of 650 mn tons in 1990 is a reasonable forecast.

GAS

The Soviet Union fulfilled its Tenth FYP target for gas production of 435 bn cu m, with Tyumen oblast producing 156 bn compared with the plan target of 155 bn. Gas is far cheaper to produce than oil with well-yields (in terms of standard fuel) generally being much higher and the recovery rate far greater. But gas costs five times as much to transport, with the throughput of an oil pipeline being three times that of a gas pipeline of similar size. The biggest constraint on gas production is therefore the availability of large diameter pipe of 1,420 mm, which the Russians have only just started producing, and they have purchased large quantities from Japan and West Germany during the last decade.

The long-term plans for gas production have been raised at the expense of those for oil. In 1978, it was predicted by the secretary of Tyumen obkom and former head of Tyumengazprom, Y.Altunin that output by Western Siberia would rise by 135 bn cu m a year during the Eleventh FYP.[1] This would give a target of 290 bn cu m for 1985. However, the official FYP target was set at 330 to 370 bn cu m at the 26th Party Congress, and this was concretised to 357 bn cu m by the November Plenum, or 67 bn cu m more than Altunins forecast. The oil equivalent of 67 bn cu m of gas is 56 mn tons and this nearly amounts to the 60 mn tons by which the original plan for Tyumen oil production was scaled down.

Output will also continue to rise in Central Asia, both from the large new Turkmen fields like Dauletabad-Donmez, and from the Uzbek deposits in the Karshi Steppe, but this will be mostly offset by declines elsewhere, notably in the Ukraine and North Caucasus.

In his EKO article, Altunin provides insights into an acrimonious debate that has been taking place over gasfield development strategy. His forecast does not include the exploitation of Yamburg, and in fact his estimate of Siberian

----

1 EKO 1979 No. 2

gas production in 1990 of 425 bn cu m plus casinghead gas assumes the exploitation only of Medvezhe, Urengoi, Vyngapur, Zapolyarnoe, Komsomolsk, Gubkin and Yubileinoe, and only "possibly" Yamburg. It is likely that under the new development strategy, Western Siberian gas production should reach 600 bn cu m in 1990.

Altunin accuses the Ministry of Gas of dismissing the smaller deposits as "small fry" that can be left for later, and of wanting to push on to Yamburg before a firm base has been created at Urengoi. Yamburg lies wholly in the continuous permafrost zone, and its development will require new technology, while the smaller deposits lie astride existing pipelines. He describes Yamburg as the Ministry of Gas's "little flower", and says that they even want to "perch on the end of the world" by extracting 35 bn cu m in 1985 from Kharasavei on the Yamal Peninsula. This 35 bn cu m would cost 1,500 mn roubles more than the cost of developing Urengoi.

A major feature of the compromise eventually agreed upon by the Ministry of Gas and Tyumengazprom was that the entire increase in output to 1985 would come from Urengoi, which would provide 250 bn cu m in 1985 compared with 190 under the original plan, and Yamburg would not come on-stream until after 1985. As long ago as 1979, it was announced that infrastructure construction work had begun at Yamal, but after the compromise was reached, the advance party abandoned its base, and did not return until January 1982, when it began dredging the mouth of the Nyudyamongotoepoka river. The development of the Yamal Peninsula gasfields seems to have been postponed until well after 1985.

While the EIU report offered a tentative forecast for 1990 of 750 bn cu m of gas, it is possible, given the new strategy of gas rather than oil, that production will be pushed as high as 800 bn cu m a year. The total production of hydrocarbons will therefore rise to 1,875 mn tsf in 1990 compared with the planned 1,644 mn tsf in 1985 at an annual rate of 2.67 per cent compared with the planned 3.58 per cent a year over 1981-1985.

CHAPTER 11

Current and Future Demand

The consumption of oil and gas is an aspect of this subject strangely neglected by Western writers. This is probably because no books have been written by Soviet specialists on the subject of fuel consumption by the Soviet economy as a whole, only by individual sectors of industry. Anybody seeking to tackle this subject must be prepared to sift through literally thousands of sources.

The best places to start in discussing oil and gas usage are the input-output tables for 1966 and 1972, which measure consumption by value in terms of delivered costs.[1]

The gross value of output of oil and oil products in 1972 came to 22.1 bn roubles compared with 10.2 bn in 1966. The most important customer was the oil refining sector, which purchased nearly 5.3 bn roubles of crude oil for processing. The next most important customers were transport with 2.69 bn roubles (1.52 in 1966), agriculture and forestry with 2.26 (1.30), electricity and thermal energy generation with 1.62 (0.46), the construction industry with 1.51 (0.59) and consumption by the population and by public authorities of 1.16 bn roubles (0.95).

The importance of oil as an input grew for nearly all industrial sectors between 1966 and 1972. This is partly due to the big increase in the price of oil products which took place in 1967, but in many cases there was a significant switch from coal to oil, particularly for electricity and heat generation, transport (i.e. railways), construction materials, and some chemical sectors. In some cases, mainly sectors of the chemical industry, there has been a switch from oil to gas.

The importance of gas has grown faster than that of any other fuel, and the gross value of output grew from 1.65 bn roubles in 1966 to 4.02 bn roubles in 1972. By far the most important consumer in 1972 was the electricity and thermal power sector which consumed gas to the value of 1,082 mn roubles, or 27 per cent of total gas sales by value. This compared with 444 mn roubles in 1966, although the share of total sales remained the same. The next most important users were households and public buildings, consuming 582 mm roubles of gas, mostly for heating buildings and cooking food. The ferrous and non-ferrous metals sector used 513 mn roubles of gas, and other important consumers were chemicals (395 mn roubles), construction materials (361 mn roubles) and engineering (317 mn roubles).

Energy saving is to be given a higher priority under the Eleventh FYP. The 26th Party Congress called for the saving of 160 to 170 mn tsf over 1981-1985, compared with 125 mn tsf saved over 1976-1980, and the November Plenum raised the target to 200 mn tsf.

While some sectors, like ferrous metallurgy or electricity generation, have very good records for fuel consumption by world standards, many other sectors

---

1 Narodnoe khozyaistvo 1967 and 1973. See also Tremi and Kostinsky: "Soviet Economic Prospects for the Seventies", and "Soviet Economy in a New Perspective".

are extremely profligate in their fuel usage. The head of the State Gas Nadzor of the Ministry of Gas, A. Kolotilin, concedes that between 6 and 8 per cent of Soviet gas (i.e. about 15 bn cu m a year) is wasted, due to the poor quality of industrial furnaces and their inefficient use. Gas-fired industrial furnaces are still manufactured on a on-off basis because the Ministry of Electrical Engineering has failed to organise their serial production, although it was instructed to do so several years ago. This has made it more difficult to carry out a planned programme of changes in design so as to get better fuel usages, and Kolotilin claims that some furnaces have only 33 per cent the fuel efficiency of imported models. They have inefficient burning devices, and are not usually fitted with automatic heat regulators.

The problem of poor equipment is compounded by its inefficient use, with an estimated 4.2 bn cu m of gas wasted every year for this reason. The main culprits are the ministries responsible for the petrochemical, automobile, paper and heavy engineering industries. Some 40 per cent of the enterprises investigated do not have meters for measuring their consumption of heat and steam and more than one third of the petrochemical, automobile and energy engineering enterprises do not employ boiler utilisers. A further 1.9 bn cu m a year are lost by the poor regulation of heating boilers serving the residential and municipal sectors; half of these boilers are not even equipped to measure their output of heat.

The Soviet press often carries stories on the success of particular factories in saving fuel and energy. The Magnitogorsk iron and steel plant, the Volkhov aluminium smelter, the Uralkhimmash chemical engineering plant and the Ust Kamenogorsk non-ferrous metals combine are frequently mentioned in this respect. These are all very large plants, but it is the medium and small-sized plants which arc responsible for a large share of the fuel wastage. While consumption by a large plant can be made the subject of a fuel-saving campaign, with usage monitored by every workshop and at every level of production, it is physically impossible to use these administrative methods at every small plant, and cost incentives have to be used. Accordingly, a new fuel and energy price schedule has been drawn up with big increases in the cost of oil products, gas and electricity. It was introduced on the 1st January 1982.

Of the 200 mn tsf to be saved over 1981-1985, the greater use of secondary energy resources (waste steam and heat, etc.) and the growing importance of centralised (as opposed to local) sources of thermal energy will account for 55 mn tsf, a reduction in the usage of electricity and thermal energy per unit of production will save 76 mn tsf, and 15 mn tsf of light oil products will be saved by improving the fuel efficiency of transport. This will be largely achieved by raising the share of diesel in road transport, because 25 per cent less fuel is needed by diesel engines compared with petrol engines for the same mileage.

The electricity and thermal energy generating sector has been asked to save 12 mn tsf during the Eleventh FYP period, principally by reducing the unit fuel consumption for electricity production. Ferrous metallurgy must save 8 mn tsf, with the expenditure of fuel on the production of pig iron and the more energy intensive types of rolled steel falling by 0.5 mn tsf annually. The plan to extend the share of steel produced by the oxygen converter process, and the greater use of continuous casting should provide big savings. Losses of thermal energy by the engineering, light industry and other manufacturing sectors waste an estimated 100 mn tsf a year, and this can be partly prevented through a greater use of recuperators planned for the Eleventh FYP period.

Big efforts will be made to reduce the 400 mn tsf a year spent on the provision of heating and hot water to residential and municipal buildings, partly by replacing a large number of small inefficient boilers by a smaller number of large ones, and partly by a better regulation of the amount of heat delivered to buildings. At the moment, 60 per cent of boilers producing this heat are not fitted with measuring devices. The problem is far worse with industrial boilers. Of 46 industrial ministries, only 3 (administering the electronics, machine tools, and agricultural machinery sectors) can give information on their usage of heat from industrial boilers. The other 43 do not, apparently, make any attempt to measure it, and this makes the elaboration of norms for the consumption of heat more difficult. It is estimated that 80 mn tsf a year is wasted in this way.

While the big increase in the production of measuring instruments called for by the plan may help to reduce heat (and hence fuel) consumption, it is also recognised that the price increases of 1982 and 1983 will prove an important factor.

This study will look in greater detail at the consumption of oil products and gas by the electricity and thermal energy sector, transport, chemicals, and final demand by the population and public authorities.

THE ELECTRICITY AND THERMAL ENERGY SECTOR

The Tenth FYP expected the production of electricity to grow from 1,038 bn kwh in 1975 to 1,380 bn in 1980. This target was not met, with only 1,295 bn kwh produced in 1980, and the annual growth rate over the last FYP period amounted to 4.5 per cent. Production by conventional thermal power stations rose at an average annual rate of 3.2 per cent to 1,038 bn kwh, compared with 7.9 per cent a year to 184 bn kwh at hydro stations and 21.1 per cent a year to 73 bn kwh at nuclear stations.

The Eleventh FYP has set a target of 1,555 bn kwh in 1985, implying an annual rate of growth of 3.7 per cent over 1981-1985. The production of nuclear power is planned to grow at 24.7 per cent a year to 220 bn kwh in 1985, hydro power by 4.6 per cent a year to 230 bn kwh, and power from conventional thermal stations by only 1.3 per cent a year to 1,105 bn kwh.

Table 32

Production of Electricity in the USSR (bn kwh)

	1975	1978	1979	1980	1985 Plan
Total	1,038	1,202	1,239	1,295	1,555
Thermal	885	988	1,012	1,038	1,105
hydro	126	170	172	184	230
atomic	28	44	55	73	220

Source: Narodnoe Khozyaistvo 1980, Pravda 4/2/81

During 1976-1980, the total capacity of Soviet power stations grew by 49,200 MW, which represented an absolute and relative decline in growth rates over 1971-1975 when 51,300 MW were installed, and 1966-1970 (51,100 MW). Over the Eleventh FYP period, it is planned to commission 69,000 MW of new capacity, including 23-25,000 MW of atomic and 12,300 of hydro. Perhaps 34,000 MW or so of new conventional thermal capacity will be installed, while as much as 21,000 MW of aging thermal units will be scrapped.

Table 33

Capacity of the Soviet Electricity Industry ('000 MW)

	1975	1978	1979	1980	1985 Plan
Total	217.5	245.4	255.3	266.7	315.0
Thermal	170.8	188.8	192.8	201.3	214.4
Hydro	40.5	47.5	50.0	52.3	64.6[1]
Atomic	6.2	9.1	12.5	13.1	36.0

Source: Narodnoe Khozyaistvo 1980,

1 Izvestia 21/4/81

Thermal power stations include condensing stations, heat and power plants (HPP's) and small gas or diesel-powered mobile plants used in remote areas for building sites etc. At the end of 1980, the USSR had 201,300 MW of thermal capacity, including 120,000 MW of condensing stations, 76,000 MW of HPPs and 5,300 MW of mobile stations.

The electricity supply industry also produces a large amount of thermal energy. In 1980, HPPs produced 1,090 mn G-cals, which accounted for 48.2 per cent of all thermal energy produced in the USSR by centralised sources (i.e. power stations, industrial and large municipal boilers). The Tenth FYP target for 1980 of 1,250 mn G-cals was significantly underfulfilled.

The output of thermal energy from all sources amounted to 3,130 mn G-cals in 1980, compared with 1,900 mn in 1970, an increase of 65 per cent. Output by centralised sources rose from 1,297 mn G-cals to 2,260 mn, or by 74.3 per cent, and this reflects the relative stagnation in output from small, comparatively inefficient local sources. The plan for 1985 calls for the production of 2,670 mn G-cals from centralised sources, an increase of 18.1 per cent over 1980.

According to the input-output tables of Narodnoe Khozyaistvo, the electricity and thermal energy sector used 5,334 mn roubles worth of fuel in 1972, of which coal accounted for 44.6 per cent (down from 58.1 per cent in 1966), fuel oil 29.7 per cent (up from 17 per cent), gas 20.3 per cent (16.7 per cent), peat 3.4 per cent (6.6 per cent), oil shale 1.3 per cent (same) and crude oil 0.7 per cent (1.4 per cent). It should be remembered that these costs are determined by the price of fuel to the power stations and boilers and therefore include transport and distribution costs as well as turnover taxes on the fuel. The change in the share of individual fuels in the total value of fuel consumption depends not only on the changing fuel balance in physical terms, but also on the size of the price increases of 1967.

Between 1975 and 1980, the share of fuel oil in the total fuel requirements of Soviet power stations arose from 29.4 per cent to over 35 per cent, but is planned to fall back to less than 26 per cent in 1985. While the share of gas rose slowly over 1976-1980 from 22 to 24 per cent, it should grow rapidly to nearly 32 per cent in 1985.

Table 34

Consumption of Fuel by Soviet Power Stations (mn tsf)

Coal	163.4	204.7	204.7	231.7
Oil	80.8	135.7	190.0	151.5
Gas	81.9	101.2	130.7	184.3
Others	18	18	15	17
Total	344	460	540	585

Source: estimates by the author, based on Teploenergetick No. 5, 1981

Over 722 mn tsf were required for the production of electric and thermal energy by power stations and both centralised and local boilers in 1980. This amounted to 43 per cent of all fuel consumed in the USSR. The consumption of fuel by power stations grew by 80 mn tons (17.4 per cent) over the five years 1976-1980, while consumption by other centralised and local sources of thermal energy grew by 18.2 per cent.

Table 35

Consumption of Fuel by the Electricity and Thermal Energy Sector
mn Tons of Standard Fuel

Power stations (inc HPPs)	460	515	520	540
for electricity	301	327	334	340
for thermal energy	159	187	194	200
Other centralised & local				
sources	154	162	177	182
Total consumption of fuel	614	677	705	722

Source: Narodnoe Khozyaistvo and authors estimates from data on production of electricity and thermal energy and the unit consumption of fuel.

The USSR has made considerable strides since the War in reducing the amount of fuel it requires to produce a unit of electricity. In 1945, 627 grams of standard fuel (gsf) were required for one kwh of electricity, but in 1980 only 52 per cent as much fuel was needed.

Table 36

	1975	1978	1979	1980	1981	1982*
Electricity (gsf/kwh)	340	331	330	328	327.1	324.5
Thermal energy (kgsf/G-cal)	173.6	173.1	173.0	173.0	172.9	172.9

Source: Narodnoe khozyaistvo 1980 Teploenergeticka No. 2, 1982
*Plan

The low fuel usage rate of 327 gasf/kwh is due primarily (some Western writers say entirely) to the importance of HPPs in total Soviet generating capacity, with some of these requiring less than 170 gsf/kwh for the production of electricity. But other reasons are the high share of fuel oil, gas and good-quality coal in the fuel mix, and the high share of large base-load units of 300, 500 and 800 MW, with the world's largest unit of 1,200 MW now operating at Kostroma. Some condensing stations are therefore able to claim extremely good usage rates. Kostroma, for example, used 317.8 gsf/hwk over 1976-1980 compared with a plan target of 327, enabling it to save 200,000 tons of fuel oil. The gas-powered Sredneuralsk needed only 316 gsf/kwh, compared with an average 358 for all condensing stations and 267 for HPPs, in 1978.

The USSR compares very favourably with other industrial countries in respect to fuel consumption, with only the highly efficient French electricity industry requiring less fuel per kwh. If the Soviet output of 1975 had been produced with the USA's rate of consumption in 1973 (which had actually been rising for five successive years) then the USSR would have required a further 26.5 mn tons of standard fuel.

In some regions, such as Moscow, highly efficient HPPs produce most of the thermal energy and account for a large share of electricity. This enables the Mosenergo Association to achieve very good rates of fuel usage. But in other regions, such as the Eastern Ukraine, fuel usage rates are comparatively poor because many of the stations are technologically obsolete and work on low grade coal with a low heat content.

The greatest scope for reducing fuel consumption by the sector is by increasing the degree of centralisation of production of thermal energy. This is obviously difficult in small towns where demand is not sufficient to justify the construction of a large HPP. But in many large towns, heat for factors and dwellings is still provided by thousands of small individual boilers using up to 400 KgSF/G-cal compared with 173 by HPPs. As well as better fuel usage, great savings can also be made through lower capital costs, lower costs of fuel transportation and a smaller number of servicing personnel.

The trend towards greater centralisation of heat supply has been apparent for a long time. In 1965, HPPs, industrial boilers and large municipal heating boilers accounted for 60 per cent of all heat supply, in 1970 for 68 per cent and in 1980 for 75 per cent. Today, practically all the demand for steam and hot water by industrial enterprises, and 40 per cent of the demand by the population and public authorities is met by centralised sources.

The basis of government policy in this respect is the HPP, which the directives of the 24th Communist Party Congress said would achieve "the rational concentration and centralisation of production of steam and hot water for technological and heating needs, and the gradual disappearance of the small boiler". This policy conflicts with that of building atomic stations and large thermal power stations, which satisfy the demand for electricity and result in a "tendency towards the limitation on the construction of HPPs with their substitution by regional and industrial boilers". Consequently, the share of HPPs in total electricity generating capacity has fallen from 40 per cent in 1965 to 36 per cent in 1978, although the total capacity of HPPs continues to grow. Large industrial and municipal boilers are becoming progressively more important.

At the end of 1980, more than 800 towns had a centralised heat supply, with HPPs producing more than 50 per cent of their total requirements. It is planned to shut down 3,000 small and obsolete boilers in 30 towns during the Eleventh FYP period.

This has had major implications for fuel consumption; nearly all new centralised sources of thermal energy operate on natural gas, and most of the small boilers

---

1 A. Gorshkov: "Tekhniko-ekonomicheskie pokazateli teplovykh elektrostantsii" 1974 p. 212.

which are being scrapped worked on fuel oil or coal. Natural gas is the cheapest of all fuels available to Minenergo planners, and they would undoubtedly prefer to use it for all their projected thermal power stations. It has been estimated that in 1972 its delivered cost to the power industry averaged 10.4 roubles per ton of standard fuel compared with 13.3 roubles/ton of standard fuel for coal and 17.3 roubles/ton of standard fuel for fuel oil. As well as its lower delivered costs, it is easier to use; it can be piped directly into the station, and does not have to be stored like coal or fuel oil. The problem at the present time is that the supply of natural gas is unable to meet even a fraction of the demand for it and power stations come right at the bottom of the list of priorities. With the rapid expansion of industries using gas as a raw material, as well as those where the use of gas in preference to coal leads to a considerable improvement in the quality of the product, it looked unlikely only a short time ago that the share of gas in total fuel consumption by the electricity sector would rise as rapidly as in the past. However, the new policy of boosting gas production at the expense of oil as well as the failure of the coal industry to even increase output, never mind meet its plan targets, implies that the programme of converting oil and coal-fired stations to gas will be sharply accelerated.

No new oil-fired power stations are to be built in the Soviet Union, and existing stations are being converted to burn gas during the summer; this policy has the dual purpose of cutting fuel oil consumption, and simultaneously evening out the seasonal fluctuation in the demand for gas, which falls sharply in the summer. Practically all new heat and power plants are to be gas-fired, and this policy has been pursued for a number of years now for anti-pollution reasons. Their role as suppliers of thermal energy and the physical limitations on the distance over which thermal energy can be transported means that they must be built close to the centres of cities.

In the Urals, coal-fired stations will be converted to gas because the local coal-fields are declining and the cost of bringing coal from further east is becoming too high. During the 1970s, the transportation of coal by the railways rose from 425 bn ton-kms to 599 bn, accounting for 17.4 per cent of total rail transport turnover. In the future, the policy will be to use coal locally while transferring coal-users at great distances from coalfields to natural gas. The Urals region is an obvious case, with a number of power stations having to rail in low grade coal from Ekibastuz. These stations will be converted to gas delivered from Urengoi through the Urengoi-Chelyabinsk pipelines while Ekibastuz coal will be burned by the massive Ekibastuz mine-side complex currently under construction.

The use of gas is also increasing in Central Asia, where the region's largest power station, Syr Darya, serving Tashkent, was recently converted to gas which arrives through a new 400 km pipeline from Shurtan. It formerly worked on fuel oil which had to be railed 2,000 km from Pavlodar. The projected Talimardzhanskaya station of 3,200-mw will also work on natural gas from Shurtan and, over the years, the few remaining Central Asian stations using coal or fuel oil will be converted to gas.

Apart from the 4,800 MW station at Perm, designed to burn Kuzbass coal, the Russians do not plan to build any new base-load thermal power stations in the European part of the country, where additional power is to be obtained almost entirely from nuclear stations. New thermal power stations fuelled by casinghead gas at Surgut and by Natural gas at Urengoi in Western Siberia, the two large mine-side complexes at Ekibastuz in Kazakhstan and Berezovsk on the Kansk-Achinsk coalfield in Eastern Siberia. The speed at which these two projects are developed depends on the extent to which the Russians succeed in building and operating transmission lines of very high tension, i.e. 1,500 and 2,200 kv.

The Ekibastuz complex will consist of at least five power stations of 4,000 MW capacity each, and its eventual capacity may reach 25,000 MW. The first station has been producing power for customers in the Urals since April 1980 when its first 500 MW unit started up after a long delay, and its installed capacity has now reached 2,000 MW. A 1,500 kv DC transmission line stretching 2,500 kms from Ekibastuz to Tambov in central Russia and a 1,150 kv AC line to the Urals are now being built. Construction of the Ekibastuz-2 power station has begun; it will also be of 4,000 MW and will begin producing in 1983.

The latest plan for the Kansk-Achinsk complex, known by the Russians as KATEK, is to create eight large quarries producing 320 mn tons of low-grade brown coal a year. As many as ten large stations with an aggregate capacity of 64,000 MW may eventually be working here, and the first unit of the first station is planned to start up in 1984. However, KATEK is unlikely to contribute significantly to the country's energy balance before the late 1980s at the earliest.

It is clear that nuclear stations will account for an increasingly large share of total electricity output during the 1980s. This policy stems not so much from fears about the future availability of fuel, but more from the fact that nuclear power generation is very cheap now that 1,000 MW units are being installed, and it will become even cheaper when these are produced on a serial basis. The USSR currently has 16,500 MW of nuclear capacity.

The Tenth FYP target of 19,400 MW of nuclear capacity producing 80 bn kwh by 1980 was underfulfilled, with 13,120 MW installed and giving 73 bn kwh in that year. This was due to unforeseen difficulties encountered in building the 1000 MW units, which have now become standard for all nuclear stations other than small ones serving remote areas. During the Eleventh FYP period, the installation of nuclear capacity is planned to proceed twice as quickly as during the Tenth, and the plan envisages the installation of 23-25,000 MW, bringing total capacity in 1985 to 36-38,000 MW. Tentative plans for 1990 foresee a nuclear capacity of 80,000 MW, able to produce 480 bn kwh of electricity. The most crucial factor affecting the fulfillment of these targets will be how soon the Atommash plant at Volgodonsk, which will be able to produce eight units of 1,000 MW each a year can begin working at full capacity. It completed the first reactor housing in February 1981, and is currently engaged on making parts for reactors rather than the reactors themselves.

In the meantime, reactors will be produced by the Elektrosila plant at Leningrad and the Skoda Engineering Plant in Czechoslovakia. The Skoda plant has planned to build 19 Novovoronezh-type reactors of 440 mw each, and an unspecified number of 1,000 mw reactors by 1985. Perhaps half of these are intended for export to the USSR.

The USSR has thirteen nuclear stations giving power at the present time, with an aggregate capacity of 16,500 MW. These are: Leningrad (4,000 MW), Novovoronezh (2,455 MW), Chernobyl (3,000 MW), Kursk (2,000 MW), Kola (1,320 MW), Beloyarsk (900 MW), Armenia (880 MW), Rovno (880 MW), Shevchenko (350 MW), Bilibino Atomic HPP (48 MW), the Siberian plant (probably 600 MW, but location unknown as it is a military plant), Obninsk (5MW - the world's first commercial station, now engaged in research), and Dimitrovgrad (62 MW).

Work is now taking place on extensions to existing stations and the construction of new stations, which will raise the total capacity of nuclear stations by 68,440 MW to 84,940 MW. The extensions are: Chernobyl by 4,000 MW, Rovno by 4,000 MW, Kursk by 2,000 MW and Kola by 440 MW. The new stations being built are: Zaporozhe, Smolensk, Nikolaev (known as "South Ukraine"), Khmelnitskii, Blalkovo, Kalinin, Nizhnehamsk, Crimea and Rostov (all 4,000 MW each), Bashkir

Ignalina and Kostroma (6,000 MW each) and the Odessa and Minsk atomic heat
and power plants (2,000 MW each). Atomic boilers are being built at Voronezh
and Gorki. Fast breeder reactors of 800 and 1,600 MW are to be designed.
These will be based on the successful 600 MW reactor at Beloyarsk, said to
be the world's largest fast breeder, but plans to install such units during the
next five years have not been announced.

With a target of at least 23,000 MW over 1981-1985, an average 4,600 MW a year
of nuclear capacity must be installed. 1981 saw the completion of 2,880 MW,
consisting of the 4th set at Leningrad and the 3rd at Chernobyl (1,000 MW each),
and the 4th at Kola and the 2nd at Rovno (440 MW each). The average rate for
1982-1985 has therefore risen to 5,030 MW a year. This is not impossible,
especially if the three sets of 1,000 MW each which should have been commissioned
in 1981 at Kursk, Smolensk and Nikolaev are completed quickly. But the Russians
are sufficiently concerned to have set up a special committee under the Council
of Ministers, and including some of the country's most powerful figures, to
oversee the construction programme. The 1985 target may well prove impossible
to achieve if the rate of work is not improved.

Given the question mark over the fulfillment of the 1985 plan, an installed
capacity of 70,000 MW in 1990 seems likely; this would be sufficient to produce
420 bn kwh of atomic power in that year.

It is estimated that the potentially possible capacity of hydro-stations is
450,000 MW, of which 240,000 MW is technically feasible and 125,000 MW is
economically justifiable at the present time. The Tenth FYP called for total
hydro capacity to rise from 40,500 MW in 1975 to 53,600 MW in 1980, but this
was underfulfilled, with 52,300 MW installed at the end of 1980.

In 1970, Gosplan believed that total installed hydro capacity could be increased
to 100,000 mw by 1990, but 60 per cent of this would be located in Eastern
Siberia and would be dependent on the development of power lines of up to
2,200 kv which could transmit electricity westwards, or even eastwards to Japan.
Such a capacity would not now be feasible, since hydro stations take up to
20 years to build, and an installed capacity of 70,000 mw seems more reasonable,
given the size of the stations currently under construction.

At the present time, the major stations under construction are: Sayano-Shushen-
skoe (where 3,840 MW of an eventual 6,400 MW has been installed so far),
Boguchany (4,000 MW, where the dam is now being built), Bureya (2,00 MW), Kolyma
(720 MW) and Kureisk (500 MW), which are all in Siberia, and Shulbinsk (1,350
MW), Rogun (3,600), Kurpsai (800) and Baipazinsk (600 MW) in Central Asia.
In the Caucasus, Khudon (740 MW), Shamkhor (760) and Inganai (800 MW) are
being built.

In the European part of the Soviet Union, the scope for new stations is limited.
Three new stations are under construction - Nizhnekamsk on the Kama, where 624
MW of an eventual 1248 MW has been installed, Cheboksary on the upper reaches
of the Volga (224 MW from 1,404 MW) and Knestrovsk on the Dnestr (238 from 714
MW). One more large hydro station, Tashlyk (1,800 MW) on the Yuznyi Bug is to
be built. In addition, several pumped storage stations are being built. In
the Eleventh FYP it is planned to install 12 turbines at the new stations at
Zagorsk (north of Moscow), Kaisyadorys (Lithuania) and Konstantinovsk in the
Ukraine. The last station will work in conjunction with Nikolaev nuclear station
and Tashlyk hydro station. The first 200 MW units are expected to start up at
Zagorsk in 1983. During 1986-1990, a further 38 turbines with a total capacity
of 6,000 MW will be built at pumped storage stations. At the present time,

work is being carried out on the expansion of existing stations, or the building of new stations, involving 27,700 MW of new hydro capacity. Under current plans, a total installed capacity of 75,000 MW in 1990 looks probable, producing 265 bn kwh.

The development of power stations based on renewable energy sources is also continuing. The Soviet Union's first solar power station of 5 MW is being built near the site of the Crimea atomic power station at Aktash by Kneprstroi of Zaporozhe. It will consist of a computer-operated system of mirrors which move in conjunction with the sun and reflect sunlight on to a single point on a boiler. It should begin operating before 1985. The generation of power from geothermal energy should increase; at the moment, the biggest station is the 11 MW Pauzhetka on Kamchatka, but preparations are under way for the construction of a 400 MW station based on a group of 5,500 metre deep wells in the North Caucasus. However, fossil-fuel power stations will continue to produce most of the Soviet Union's power until well into the 21st century.

The rate of fuel usage should continue to fall, perhaps to 320 gsf/kwh by 1985 and 315 by 1990 because highly efficient new sets of 500 and 800 MW at baseload stations, or up to 250 MW at HPPs will be accounting for a larger share of installed capacity. It is possible to forecast that power stations will require 355 mn tsf for production of electricity and 230 mn tsf for the production of thermal energy in 1985, giving a total requirement of 585 mn tsf. This represents a growth rate of 1.6 per cent a year over 1980.

Other centralised and local sources of thermal energy may require 200 mn tsf; while their total capacity will increase by up to 5 per cent a year, the process of replacing small boilers (with rates of fuel usage of more than 400 kgsf/G-cal) by highly efficient industrial and municipal boilers will ensure that overall fuel consumption grows very slowly. The whole electric and thermal energy generating sector may require 785 mn tsf in 1985 compared with 722 mn tsf in 1980, and consumption will grow at an average annual rate of 1.7 per cent a year at the same time that Soviet fossil fuel production is planned to grow by 3.15 per cent a year. If the fuel production target is met, and the forecast of consumption by the power industry is correct, then fuel consumption by other sectors, including exports and stock-building, should be able to grow at an average annual rate of 4.0 per cent, from 1,174 mn tsf to 1,427 mn tsf in 1985.

Forecasts for 1990 must be more tentative, especially as nuclear HPPs will be accounting for a significant share of Soviet thermal energy output. A reasonable guess is that 1,850 bn kwh of electricity will be produced with 420 bn coming from nuclear stations and 265 bn from hydro stations. This leaves 1,165 bn kwh from thermal stations, or only 60 bn more than in 1985, suggesting an average annual increase of less than 1 per cent over 1986-1990. With the completion of the Ekibastuz complex and the rapid build-up of KATEK, this implies that aging power stations in the European part of the USSR, mainly coal and oil burning, will be scrapped at an unprecedented rate.

By 1990, almost all the country's thermal energy should be derived from HPPs or centralised boilers. It seems that an important change in policy is taking place, with more emphasis on large industrial and municipal boilers at the expense of heat and power plants, and the impact of this policy should become more apparent during 1986-1990. Accordingly, fuel consumption in 1990 for the production of thermal energy by power stations should be less, and by centralised boilers correspondingly more, than anticipated by the EIU report of 1980. A requirement of 630 mn tsf by power stations and 240 mn tsf by centralised boilers seems probable, giving a total of 870 mn tsf by the sector.

It is possible to surmise that by 1990, many of the old condensing stations in the European part of the country (especially coal-fired stations in the Ukraine) will have been scrapped. The importance of coal will have grown further since 1985, because the Ekibastuz complex will be providing a large part of the European USSR's needs, while the Kansk-Achinsk will be starting to cover a significant part of the electricity requirements of the Urals and Volga regions. Half of the electricity produced in the Ukraine will come from atomic stations. The share of gas will also be rising as a fuel for HPPs and boilers. The volume of fuel oil burned in power stations will be falling in both relative and absolute terms, as the steady rise in secondary refining capacity permits more and more fuel oil to be refined still further. By 1990, super-high-tension power lines should be beginning to bring power to Central Russia from Siberian hydro-stations.

TRANSPORT

In spite of the recent growth of road transport, the railways still account for 70 per cent of total freight turnover in the Soviet Union. In 1980, 3,728 mn tons of freight were carried an average of 922 kms, giving a freight turnover of 3,439 bn ton-kms. Poor quality roads and immense distances between major cities make road transport for long hauls uneconomical, although the growing sophistication and diversification of the Soviet economy is leading to a greater importance of short-distance hauls. Consequently, freight turnover by road transport has grown at an annual rate of 7.0 per cent during the 1970s compared with only 3.3 per cent for rail. In 1980, Yerries carried 24,100 mn tons of freight for an average distance of 17.9 km.

Table 37

Freight Turnover and Volume by Different Types of Transport

	1975	1978	1979	1980
Turnover (bn ton-kms)	4,532	4,896	4,843	4,964
rail	3,236	3,429	3,349	3,439
road	338	396	410	432
sea	736	828	858	848
river	222	244	233	245
Volume (mn tons)	25,200	27,700	27,850	28,600
rail	3,621	3,776	3,688	3,728
road	20,900	23,200	23,400	24,100
sea	200	229	227	228
river	476	546	537	568

Source: Narodnoe Khozyaistvo 1980

In complete contrast with other industrially developed countries, the Soviet rail network is steadily growing in size, at a rate of 660 km a year during the 1970s to 141,800 kms in 1980. This is primarily due to major projects such as the BAM railway and the system being created in the Western Siberian oil and gas regions.

The degree of track utilisation is also growing, and reports of severe congestion in certain areas appear from time to time in the Soviet press. In 1965, the Soviet railways carried 18,381 tons of freight for each kilometre of track, and by 1980 this figure had grown to 26,290 tons. This is in spite of the expansion of the oil pipeline system helping to relieve congestion by carrying loads which would otherwise have gone by rail.

Other factors relieving congestion have centred round the change from steam to diesel and electric traction. Steam trains have virtually disappeared now, and electric traction is growing in importance. Electrified lines include those most heavily used; in 1980, 31 per cent of the total length of track was electrified but 55 per cent of all freight was carried by electric trains.

The importance of electric compared with diesel traction obviously has a major influence on the level of direct fuel consumption by the railways. Although the degree of electrification is far higher than in other industrial countries, it can grow further. With the rapid growth of supply of electricity from atomic stations in the European USSR and from hydro stations in Asiatic USSR, it may be that the advantages of electrification will outweigh the greater costs. Electric trains can pull heavier loads than diesel trains and can travel much faster, with the average daily run of an electric train amounting to 524 km in 1978 compared with 455 by diesel train.

More than 90 per cent of freight turnover by sea transport is accounted for by foreign trade. Consequently, the average journey is very great at 3,720 km in 1980, and the turnover is out of all proportion to the volume of freight, which accounts for only 0.8 per cent of the total. The impetus to long-distance foreign trade began in the late 1950s with the growth of Soviet trade with developing countries. Since then, the Soviet merchant fleet has grown rapidly at an annual rate of 10 per cent to over 24 mn gross registered tons in 1980 from only 3.4 mn in 1960.

The haulage of freight by the inland waterway system has grown more slowly than that by other types of transport. This can be attributed to the stagnation in the total length of the waterway network.

Road transport has been growing thanks to the improvement in the condition of Soviet roads and the growth in the size and capacity of the nation's lorry fleet. Although all the principal routes are now hard-surface, with motorways being built in heavily populated regions, the share of hard-surface roads in the total length of roads is continuing to grow under a long-term plan to give a hard surface (usually compacted gravel) to rural roads. In 1980, 72 percent of the USSR's 1.02 kms of general purpose roads were hard surface compared with only 44 per cent in 1970. Hard surface roads give fuel savings of up to 30 per cent over soft surface roads.

There is no official data on the size of the Soviet lorry fleet, although it can be estimated that it amounted to 7.1 mn by the end of 1980 compared with 3.8 mn in 1970. The lorries are also getting bigger; the average lorry produced in 1970 had a load capacity of 4.3 tons while by 1980 it had risen to 5.0 tons. The commissioning of the Kama Lorry Plant helped to raise the average load capacity from 4.5 tons in 1978 to 5 in 1980. The quality of engines is improving, and better lorries and roads have permitted an improvement in the rate of fuel usage of 40 per cent over the last 20 years.

In spite of the increasing level of private car ownership, the use of public transport is still continuing to grow. In 1980, the average Soviet citizen made 256 journeys by public transport compared with 238 in 1975. These consisted of 158 by bus (143 in 1975) 5 by public taxi (7), 34 by trolleybus (31), 31 by tram (32), 13 by train (13), 15 by metro (12) and one by ship or riverboat. Private car ownership has been growing rapidly in recent years. Since 1950,

the USSR has produced over 16 mn cars, of which 25 per cent have been exported.
With the construction of the Tolyatti car plant, annual output has grown from
344,000 in 1970 to 1,327,000 in 1980. It can be estimated that there were
9.5 mn cars in the USSR at the end of 1981 compared with 1.8 mn in 1970.

The share of transport in the consumption of fuel and energy has fallen from
19 per cent to 10 per cent in the last 20 years. Between 1959 and 1966, the
value of fuel consumed by the transport industry (excluding agricultural vehicles
and private cars) did not increase, and its share of the total costs incurred
by the transport sector fell from 15.6 per cent to 9.3 per cent. This was due
to the conversion of the railways from steam to diesel traction, and hence from
expensive coal to cheap oil. During the period 1966-1972, the conversion of
oil continued, and the oil price rise of 1967 resulted in a big increase of
fuel costs to 2,900 mn roubles. Since 1972, the use of coal has almost disappeared, with the value of oil consumption rising fairly slowly due to the
process of conversion from diesel to electric traction now taking place.
The value of electricity used by the sector (for urban transport as well as
electrified railways) has grown from 150 mn roubles in 1959 to 458 mn in 1966
and 820 mn in 1972.

The consumption of fuel by the railways has actually halved between 1960 and 1975
from 60.2 mn tsf to 30.2 mn tsf. During this period, consumption of coal fell
from 42.8 mn tsf to practically nothing, and fuel oil from 10.6 to 2.1 mn tsf,
while the consumption of diesel fuel rose from 3.8 to 12.8 mn tsf. By 1975
only 4.87 tsf per mn ton km were needed compared with 20.8 in 1960. Any further
reduction in unit fuel consumption will depend on technical improvements to
engines, and increasing the speed and size of trains. This is being attempted;
trains 1½ km long, for example, are now being use to ship coal from the
Kuzbass to Moscow. The Voroshilovgrad Locomotive Plant, which makes 96 per
cent of Soviet diesel trains, has begun the serial output of 8,000 HP engines,
and the construction of a 9,000 HP prototype is taking place.

Diesel is tending to replace petrol as the major road transport fuel, because
diesel drive permits fuel savings of up to 25 per cent over petrol. Official
data have not been provided either for the production of diesel or for its
consumption by different types of transport except the railways. It can be
estimated that 120 mn tons of diesel were produced in 1980, of which 20 mn
tons were exported leaving 100 mn tons for internal use. Rail freight transport
required 10.5 mn tons (the same as in 1975), and the rest was used by passenger
railways, road, river and sea transport enterprises, and also by farms for their
tractors (now using diesel rather than kerosene), combine harvesters and trucks.
Large volumes were used to power stationary motors of different types by virtually all sectors of industry.

In 1980, it can be estimated that 68 mn tons of petrol were produced, of which
10 mn tons were exported, leaving 58 mn tons for domestic consumption. The
8.5 mn cars on Soviet roads in 1980 can be estimated to have used 16 mn tons
of petrol; this is more than the same number of cars would use in Western
Europe because the Russian cars include a disproportionately large number of
taxis which travel huge distances each year. The rest of the petrol was
used by vans, lorries, buses, some agricultural vehicles and the military.

Riverboats and ocean-going ships used comparatively small volumes of fuel in 1980,
possibly 15 mn tons. Fuel usage by ships has improved by 1.8 times over the
last 20 years and by riverboats 2.4 times.

According to the Eleventh FYP, freight turnover on the railways will increase
by 14 to 15 per cent to 3,950 bn ton-kms in 1985. This is a surprisingly small

increase in view of the planned rise in industrial production, and can be explained by the policy of transferring the transport of oil and oil products to pipelines, and by the increasing localisation of coal consumption as the natural gas pipeline network is extended. The Russians plan to slow down the increase in the volume of coal being hauled long distances from the Donbass and Kuzbass to Leningrad and the western republics, and convert consumers in those regions to gas. Nevertheless, in spite of these measures, the annual volume of hard coal carried by rail is expected to grow by 62 mn tons a year from 732 mn in 1980, and the tonnage of oil loads will rise by 35 mn tons from 423 mn tons.

It can be expected that the average length of haul might fall slightly to 910 kms in 1985, with the railways hauling a total tonnage of perhaps 4,350 mn tons, or 16 per cent more than in 1980. The share of electrified track will rise to 34.2 per cent of the total by 1985, with 6,000 kms of track to be electrified over 1981-1985 while the total length of the Soviet rail network grows by 3,600 kms. The share of freight carried by electric trains is likely to rise to 65 per cent according to a speech by Minister of Communications Pavlovskii at the 26th Party Congress, thereby helping to achieve a saving of 7 mn tons of diesel fuel to be effected by freight and passenger rail transport. The accomplishment of technical measures, such as running bigger and more powerful trains, should provide further savings of 3 bn kwh of electricity and 1½ tons of diesel a year.

According to Gosplan's Institute of Complex Transport Problems, technical improvements to transport can reduce the rate of fuel usage by 10 per cent for rail transport, 8 per cent for sea transport, 7 per cent for riverboats, and 22 per cent for lorries. During the period 1981-1985, a 33 per cent increase is planned for the freight turnover of the road transport industry, from 432 bn ton-kms to 575 bn. This implies an increase in tonnage carried from 24.1 bn tons to 28.8 bn, assuming an increase in the average journey from 18 kms to about 20 kms.

Road transport, with a fuel usage rate 19 times that of the railways, accounts for half of all fuel consumed in the transport sector. The trend towards diesel rather than petrol driven lorries is to be accelerated, and this will lead to a considerable growth in diesel consumption. It can be estimated that it will grow to 101 mn tons in 1985 by all sectors of transport, including agricultural vehicles, and perhaps 130 mn tons by the economy as a whole.

The future development of passenger transport must be closely related to the increase in the stock of passenger cars. This has not yet affected the growth in the utilisation of public transport. The annual number of journeys made by the average Soviet citizen grew by 4.0 per cent a year over 1970-1980 compared with 5.3 per cent a year over 1965-1970. But with the stock of private cars likely to rise to over 13 mn by 1985, it is inevitable that this will effect public transport, although to what extent is difficult to forecast.

The Russians decided many years ago that metros were the quickest, cleanest and cheapest method of urban transportation. More recently, another attribute has acquired greater significance - they can save the country a substantial volume of diesel and petrol by attracting passengers away from buses and private cars with comparatively low rates of fuel efficiency.

Accordingly, the programme to build metros in all major cities has been stepped up. By the end of 1980, the USSR had built metros in seven cities with a total

---

1 Economist Intelligence Unit: Quarterly Energy Report, Soviet Union & Eastern Europe, 1981 No. 2. p. 43.

population of over 20 mn - Moscow, Leningrad, Kiev, Kharkov, Baku, Tbilisi and Tashkent. Between them, they had 344 km of track, an increase of 70 km over 1975, and they carried 3.8 bn passengers. With a rate of usage of 11 mn passengers per km of track a year, they are the most intensively used metros in the world.

During the period 1981-85, another 112 km of track is to be built, including 30 km in Moscow. In March 1981, the Yerevan metro began operating with the first stage of 7.6 km and three stations now in use, and by 1985 it should be 12 km long with five stations. During the rest of 1981, new lines will be opened in Leningrad and Kiev, with construction of the second stage at Tashkent continuing, and tunnelling work for the third line to begin at Kiev.

Before the end of the current plan period, the Minsk, Gorki and Novosibirsk metros should have started up. Construction of a further nine metros will begin or continue at Dnepropetrovsk, Kuibyshev, Sverdlovsk, Alma Ata, Riga, Omsk, Rostov, Perm and Chelyabinsk. By 1985, metros in 20 cities with an estimated total population of 37.5 mn will either be operating or under construction. It can be estimated that by 1985 they will be carrying 4.8 bn passengers, and by 1990, 6.1 bn. The same number of people travelling by private car would require an estimated 1.3 to 1.4 mn tons of petrol.

However, many large Soviet cities will still be without metros in 1990, including Volgograd, Donetsk, Kazan, Odessa and Ufa, all of which will have populations exceeding 1 mn in that year. Here, the emphasis will be on trolley-buses which are slower and more unsightly than metros, but which also enable large savings in oil consumption to be made. The tram services are also to be extended, not by increasing the number of trams, but by increasing speed and unit capacity, with high speed trams now operating in Kiev. Volgograd will have an underground tram system.

It is possible that, in spite of an estimated increase of 26 mn in the urban population of the USSR over the next ten years, the growth in demand for oil products for urban transport purposes can be kept to less than 1.5 per cent a year. This depends, however, on the successful fulfillment of ambitious plans for the construction of nuclear power stations to provide the power for the new urban transport networks.

With the metro being by far the fastest form of urban transport, extended car ownership should have little effect on it. Moreover, it is the most environmentally inoffensive form of transport, and as considerable capital investments have to be recouped, it is likely that efforts will be made to attract as many passengers as possible. Trams are already declining in popularity due to their slow speed, and most of the effect of mass car ownership will be felt by the buses, trolley buses and taxis. The very low fares charged for travelling by public transport mean that cars will probably be used mostly for recreation and holiday purposes.

These forecasts assume that car production will grow from its present level of 1.3 mn/year to 1.5 mn by 1985, due to the minor expansion of the Gorki and Zaporozhe car plants. The doubling up of the Tolyatti car plant may be announced shortly, as Soviet planners are preparing for the population of Tolyatti to grow from 530,000 to 1 mn by 2000. Not only has pressure for car ownership been building up among the Soviet population, but the Lada cars produced at Tolyatti have become a major Western currency earner. Exports to the UK (23,000 in 1979) should rise substantially during the next few years, and the Satra Corporation of New York expects to be importing 50,000 Ladas a year by 1985. The new Tolyatti plant should begin producing about 1985, and total car production should build up to 2.2 mn a year by 1990.

By 1985, 13 mn Soviet cars should be using 24 mn tons of petrol a year, and 70 mn tons will be consumed by the Soviet economy as a whole. Although this will represent only 13 per cent of refinery throughput compared with 50 per cent in the USA, the Russians have been making strenuous efforts to reduce the consumption of petrol by price increases and by developing substitutes.

For several years now, lorries, buses and taxis have been running on liquified natural gas on an experimental basis in Moscow, Leningrad, Kiev and Tashkent. The higher octane rating of over 100 gives an improved compression ratio, cuts exhaust fumes by up to seven times, and prolongs the life of lubricants, sparking plugs, etc. Other experiments have involved the use of a petrol-hydrogen mixture. Microbuses tested in Moscow have need only 30 per cent of their normal petrol requirements, while three taxis on test in Kharkov made petrol savings of 40 per cent.

CHEMICALS

Compared with other industrialised countries, the USSR has a small chemical industry in spite of rapid development in recent years. In 1965, total output by the chemical and petro-chemical sectors amounted to only 4.7 per cent of total industrial output, although by 1980 this share had risen to 7 per cent. There are two main reasons for this. Firstly, there has been a tendency to rely on traditional materials such as steel and wood rather than develop new alternatives like plastics. Secondly, some sectors of the industry involve highly complex and sophisticated technological processes, and it is only recently that employment of these processes has begun to take place.

Growth in the production of basic chemicals is a precondition for an expansion of output of end products such as plastics, synthetic fibres etc. The output of plastics in the USSR, while growing rapidly, is still only a fraction of that of the USA. The pressing need for the USSR to raise its output of plastics is illustrated by the fact that, during the 1960s, savings obtained by using plastics instead of traditional materials amounted to 3 bn roubles. Polyester and PBC are used to replace wood, polyurethane and vinyl are replacing leather, acrylics and polystyrene are used instead of glass, and high-impact thermosetting plastics are replacing metals. In 1980, 3.6 mn tons of plastics were produced, and the Tenth FYP target of 5.7 mn tons was substantially underfulfilled. The Eleventh FYP set a target for 1985 of 6.1 mn tons, and in 1981, production grew by 12.6 per cent to 4.1 mn tons. Over 1982-1985, output must rise at an average annual rate of 10.4 per cent for the 1985 target to be met.

The production of plastics must grow by 67.6 per cent during the current FYP period, while that of polyolefins must rise by 120 per cent, polystyrene by 130 per cent and PVC by 60 per cent. The capacity for the production of poly-carbonates must be increased by 70 per cent, mainly from a new plant which has just started up in Dzerzhinsk, and polyamides by 200 per cent from a new plant at Orokhovo-Zuevo. This will overtake the Chernovskoe plant as the largest Soviet producer of polyamides.

During the 1950s, the raw material base of the plastics industry consisted basically of phenol, formalin, methanol, benzene and camphor, i.e. products derived from coking. As recently as 1958, the coking sector accounted for 70 per cent of raw materials, plants and minerals 19 per cent and hydrocarbons only 11 per cent. In the future, the share of hydrocarbons is expected to reach nearly 90 per cent.

Natural gas is becoming an increasingly important material for plastics, especially carbamide resins, aminoplasts, polyformaldehyde etc. It is estimated that one ton of plastics can be obtained from 10,000 cu m of gas, or from 2½ tons of casinghead gas.

At the moment, the production of plastics is concentrated in the Volga region, but it has been suggested that Western Siberia should eventually account for 35 to 50 per cent of total Soviet output, although polymerised plastics should come last in order of priority after synthetic rubber and synthetic fibres. This aim will be achieved by 1985 when the Tomsk and Tobolsk plants will be producing in addition to those at Omsk and Ussolvy-Zima.

Most new petrochemical plants have been bought from the West. Since 1975, the commissioning of new plants producing plastics or materials for plastics has included:

* Ethylene and propylene: four Soviet designed plants of capacity 300,000 tons a year each, at Nizhnekamsk, Lisichansk, Gorki and Angarsk, a facility producing 450,000 tons a year at Kazan, built by Toyo of Japan, and a 250,000 tons a year plant at the Prikumsk complex in Budennyi, built by Linde of West Germany. Linde has also built a 125,000 tons a year propylene plant at Prikumsk, and the Soviet-built ethylene plants can also produce propylene - they are designated "EP-300", with the prefix standing for ethylene-propylene.

* Benzene: new plants to start up since 1975 include:

Kazan	610,000 tons/year built by Toyo	
Prikumsk	100,000	Linde
Krasnodar	150,000	Chiyoda
Mozyr	110,000	Asahi
Ufa	120,000	Eurotecnica

In 1980, the USSR produced 1.8 mn tons of benzene, compared with 1.43 mn tons in 1975. Output should grow to 2.3 mn tons in 1985 with the commissioning of new plants including:

Ufa	125,000 tons/year built by Technip and Procofrance	
Omsk	125,000	Technip and Procofrance
Novopolotsk	125,000	Asahi
Ryazan	125,000	Asahi

* Xylene: the production of xylene is concentrated at three refineries in the USSR.

Since 1975, the following plants have started up:

Para-xylene:	Kirishi	60,000 tons a year built by Kawasaki	
	Omsk	165,000	Eurotecnica
	Ufa	165,000	Eurotecnica
Ortho-xylene:	Omsk	165,000	Eurotecnica
	Ufa	165,000	Eurotecnica

In 1979, the Russians produced 435,000 tons of xylene, a rise of 33 per cent over the 327,000 tons of 1975. By 1985, output should rise to over 700,000 tons a year.

Vinyl chloride: new plants include:

Kalush	250,000 tons a year built by Uhde	
Zima	270,000	Uhde

The Zime plant is to be doubled up to 540,000 tons a year by Uhde, and the production of vinyl chloride in the Soviet Union should grow from 0.47 mn tons in 1980 to 0.70 mn tons in 1985.

Styrene: One large styrene plant has been built since 1975. It has a capacity of 300,000 tons a year, is located at Shevchenko on the Caspian Sea, and was built under a compensation agreement by Litwin of France.

Methanol: The Soviet Union currently produces over 1.8 mn tons a year, and this should grow to 3 mn tons a year by 1985 after two plants of 750,000 tons a year each are completed at Gubakha and Tomsk by the Davy Corporation.

Polyethylene: New plants include:

Low density, Kazan	120,000 tons a year, built by Salzgitter	
Severodonetck	240,000 (2 plants)	Salzgitter
High density, Prikumsk	200,000	John Brown
Kazan	200,000	John Brown

By 1979, total polyethylene output had grown to 582,000 tons a year compared with 420,000 tons in 1975. The recently commissioned plants should take it over 700,000 tons a year by 1985.

Polypropylene: There are only two plants producing polypropylene in the Soviet Union, both of which were built by Tecnimont. A plant of 30,000 tons a year was built at Gurev several years ago, and the other, with a capacity of 100,000 tons a year, came on-stream in 1981 at Tomsk.

Polystyrene: Recently completed plants include:

Dneprodzerzhinsk	100,000 tons a year, Soviet built	
Shevchenko	200,000 tons a year, built by Litwin	
Omsk	100,000	Litwin

Polyvinylchloride: Output grew slowly from 334,000 tons in 1975 to 393,000 tons in 1979, but should rise rapidly when the 250,000 tons a year facility installed in 1981 at Zima by Klockner reaches its rated capacity.

The development of chemical fibres has gradually eroded the share of natural fibres in total output, from 85 per cent in 1960 to 50 per cent today. This proportion is still higher than in the USA, and it means that the USSR is devoting an unnecessarily high volume of resources to sheep-rearing and cotton-growing.

The production of chemical fibres has grown slowly from 955,000 tons in 1975 to 1,176,000 in 1980. A target of 1.6 mn tons has been set for 1985, but production in 1981 did not rise much beyond 1.2 mn tons.

In 1959, only 7 per cent of the raw materials for chemical fibres were petro-chemicals. By 1970 this share had risen to 60 per cent, and this dramatic change is due to the greater availability of casinghead gas. One ton of synthetic fibre can be obtained from 50,000 cu m of natural gas, and enough synthetic resins can be obtained from one ton of casinghead gas to produce 3,000 sq m of synthetic fibre. According to Bobrovskii[1] "practically all synthetic fibres will be produced on the basis of benzene, xylene and other hydrocarbons derived from petroleum" in the future.

The biggest Soviet fibre plant is located at Mogilev in Belorussia. Built by Krupp-Kopper from a Dynamit N process, it has a capacity of 200,000 tons a year of DMT, and is currently being extended by 120,000 tons a year. An acrylonitrile plant of 150,000 tons a year was built under a compensation agreement by Tecnimont at Saratov, and Snia Viscosa has built a 80,000 tons a year caprolactam plant at Chirchik. However, most of the Soviet Union's 330,000 tons of caprolactam produced in 1979 came from domestically built plants, mainly at Dzerzhinsk and Grodno, where the 50,000 tons a year plant was doubled up in 1981.

The fastest growing sector of the chemical industry has been __fertilizers,__ for which ammonia is an important input. The production of ammonia has grown at an annual rate of 12 per cent since 1965 to 15.8 mn tons in 1980, and the Tenth FYP target of 22.8 mn tons for 1980 was considerably underfulfilled. Output will probably rise to 27 mn tons in 1985, with a significant share of the increase going for export. More than 90 per cent of ammonia is produced from natural and casinghead gas, with a further 5 per cent from coking gas. Coke once accounted for a significant share of raw materials, but gas has proved much cheaper. The transfer from coke to natural gas at Novomskovsk, for example, reduced unit costs by 45 per cent.

The use of coking gas is still economical in some regions. The combination of a nitrogenous fertiliser plant with a metallurgical and coking plant can reduce the unit cost of fertiliser by 5 per cent, and for a plant with a capacity of 100,000 tons of ammonia a year, a saving of 15 mn roubles a year can be effected. The best example is Novolipetsk nitrogenous fertiliser plant.

Natural gas was first used for the production of ammonia at Novomoskovsk in 1958, and by 1960 it accounted for 16.3 per cent of Soviet ammonia production, with coking gas used for 32.1 per cent, coke and coal 32 per cent and other sources of hydrogen, 19.6 per cent.

Table 38

USSR: Ammonia Production by Source of Hydrogen
('000 tons)

	1960	1970	1975	1980
Natural & casinghead gas	226	5,522	9,550	14,600
Coke oven gas	444	1,085	1,404	850
Coke or coal	443	793	684	240
Others	271	238	360	160
Total	1,384	7,638	11,998	15,850

Soviet ammonia capacity is planned to grow by nearly 8 mn tons over the period 1981-1985. The newly created Ministry of Fertiliser Production has drawn up plans for 13 new ammonia plants, each with a capacity of 450,000 tons a year in addition to the four plants currently under construction at Angarsk, Odessa, Gorlovka and Perm. In all, 17 such plants will be built with Western assistance, in addition to smaller plants of 200,000 tons a year at Rustavi and Kemerovo

1 P.A. Bobrovskii, Narodnokhozyaistvennaya effektivnost khimizatsii proizvodstva, Moscow 1968 p. 93.

156

using Soviet equipment, and this compares with 20 which came on-stream during 1976-1980.

These were located at Cherpovets, Cherkassy (2), Dorogobuzh, Tolyatti (4), making 6 at the complex with an annual output of 2.7 mn tons), Dneprodzerzhinsk, Gorlovka (2), Grodno, Novomoskovsk (2), Odessa (2), Rossosh and Novgorod (3). They were mostly built by Toyo using a Kellogg process, except those at Tolyatti which were built by Chemico with a Benfield licence, and the Gorlovka plants, built by Greusot Loire with a Kellogg process.

The Tolyatti complex is said to be the world's largest, and is likely to be extended during the next five years. It is fed by natural gas from the Urengoi-Chelyabinsk-Petrovsk pipeline, and more gas will be available from the new series of pipelines following the Urengoi-Nizhnyaya Tura-Petrovsk corridor. The growth of ammonia production at Tolyatti will be assisted by the construction of a second string of the Tolyatti-Gorlovka-Odessa ammonia pipeline, the first string of which was completed in 1980. Running for 2,450 kms with 400 shut-off valves, it carries 2.5 mn tons a year from the Tolyatti "Azot" plant. Up to 0.3 mn tons a year will be distributed to farms along its route from 30 distribution points, and 2.2 mn tons will be exported annually from the Yuzhnyi terminal near Odessa. As the second string will be of similar capacity, it is likely that half of the Soviet Union's new ammonia capacity over 1981-1985 will be built at Tolyatti.

The Novgorod plant also exports ammonia from the Baltic port of Ventspils, with Denmark and the Netherlands taking 60,000 tons a month.

During the last 20 years, the consumption of ammonia per ton of fertiliser has fallen from around 1,000 to 855 cu m per ton. The volume of gas required by the sector has therefore risen from 0.2 bn cu m to 12.5 bn cu m in 1980. It should rise to about 21.8 bn cu m in 1985, and as much as 33.6 bn in 1990.

Soviet urea is produced in equipment manufactured and installed by Chemoprojekt of Czechoslovakia using a Stamicarbon process. These plants have a capacity of 350,000 tons a year, and three were installed during 1976-1980 at Cherkassy and Dneprodzerzhinsk (2). Six more are under construction at Angarsk, Berezniki, Fergana, Grodno, Salavat and Severodonetsk. The Russians also build urea plants of capacity 450,000 tons a year each with equipment imported from Tecnimont. Two of these are operating at Berezhiki and Gorlovka, and one more is under construction at Kemerovo. Snamprogetti have built a 450,000 tons a year plant at Novomoskovsk and another 3 at Tolyatti, and Toyo have built a unit of similar size at Perm.

Fertilisers are a recent industry in the USSR, and the production of only 13.9 mn tons in 1960 illustrates the extent to which agriculture was neglected until then. Since 1960, the production of fertilisers has grown at an average rate of 10.6 per cent a year, and the share of nitrogenous fertilisers has risen from 35 per cent to 48 per cent.

Table 39

Total output	55.4	90.2	98.0	94.5	103.9
Nitrogenous fertiliser	26.4	41.6	45.4	44.6	49.9
Nitrogenous as % of the total	47.7	46.1	46.3	47.2	48.0

Source: Narodnoe Khozyaistvo

Very little fuel is necessary for the production of other types of fertiliser.

The USSR does not publish statistics on the output of synthetic rubber, but it can be estimated that 1.8 mn tons were produced in 1980. It has been argued that Siberia should eventually produce 35 to 50 per cent of total Soviet output, and this is likely to be the case when the synthetic rubber shops of the Tomsk and Tobolsk complexes come on-stream. The continued expansion of output from the Volga plants has been criticised, but the Volga region remains the largest producer of synthetic rubber, with output doubling during 1971-1975. In 1974, the Volga region produced 49 per cent of all Soviet synthetic rubber.

In 1970, synthetic rubber was obtained from the following monomers: butadiene (45 per cent), isopreme (30 per cent), chloroprene (13 per cent) and isobutylene (12 per cent). Ethylene, originally used for the production of intermediate styrene, is now being used together with Propylene in the production of special purpose EPDM rubber. Propylene is also used for the production of nitrile rubber. The cost of the final product depends largely on the type of starting material and the method of processing. The share of different materials has changed dramatically during the 1960s, with that of ethyl alcohol falling from 95 per cent in 1960 to 16 per cent in 1970. By now, its use should have practically ceased with pentane, n-butane, and isobutane being the most important materials.

Synthetic rubber was first produced in 1932 at Yaroslavl from ethyl alcohol. As early as 1950, 67 per cent of all rubber used in the tyre industry was synthetic, and natural rubber has now virtually gone out of use in the USSR.

The cost of raw materials amounts to 78 per cent of the cost of synthetic rubber and between two and four tons of materials per ton of rubber are required. Thus the importance of using efficient materials like casinghead gas (400 kg of rubber can be obtained from one ton of casinghead gas) can be readily seen. In 1959, petrochemicals accounted for 20 to 25 per cent of all materials used by the synthetic rubber sector; by 1970 this share had grown to 96 per cent.

The major problem in forecasting the future consumption of fuel by the chemical industry is that of forecasting the future output of the most important chemical products. The plan target for plastics, for example, was greatly underfulfilled during the Ninth FYP, and that for the Tenth FYP was underfulfilled to an even greater extent. This is not because the targets are particularly ambitious. The plan for 1976-1980 was to increase output from 2.84 mn tons to 5.7 mn tons; during 1961-1965, the USA's plastics industry grew from 2.85 to 5.3 mn tons a year. The reason for the underfulfillment is that all new chemical production is expected to come from a small number of very large plants. In all cases, the construction of these plants is years behind schedule, and in some cases the construction of the refineries producing the chemical feedstocks (e.g. Mazheikiai in Lithuania) suffered lengthy delays.

At the Tomsk petrochemical combine, a polypropylene facility with a capacity of 100,000 tons a year started up in 1981. In 1982, a methanol plant of 750,000 tons a year will start producing, and other plants will start up at the rate of one a year thereafter, producing ethylene, polyethylene, carbamide resin, formalin, polyformaldehydes and plastic consumer goods. When the Tobolsk combine, also in Western Siberia, becomes operational, the production of synthetic rubber will increase sharply, and car tyres, rubber articles, modern construction materials, fertilisers, paints, photographic film and consumer goods will be produced. Construction has been proceeding slowly, however; this is blamed on a faulty planning decision several years ago. It was thought that under the initial plans, the plant would be working at only 60 per cent capacity by 1990, and so it was decided to build only one instead of the original two gas fractionating plants. Siberian scientists disagreed, saying that there would be more than enough raw material for two plants. Only recently has their advice been followed, and the result is that the construction of the plant is now years behind schedule. It was planned to produce the first non-stable benzine in 1979 and the first monomers in 1981, but the plant is still in its initial stages of construction.

The pace of construction work is starting to quicken again, now that five rectification columns 90 metres high and weighing 700 tons each have arrived from the Petrozavodsk engineering plant. They will form the heart of the first fractionating plant, and produce starting materials for synthetic rubber.

The Nizhnekamsk Combine continues to grow in size and importance. In 1980, the second isoprene rubber plant started up, making Nizhnekamsk the country's largest producer of isoprene rubber, and a month later the USSR's largest ethylene oxide plant began producing materials for the plastics, synthetic fibres, detergents and paints sectors. Twelve of Nizhnekamsk's plants are now operating, and three more will start up shortly.

The fourth major complex in this series is the Prikumsk Chemical Combine at Budennyi in the North Caucasus region, which began producing ethylene from its 250,000 tons a year ethylene facility in 1980. This will permit the production of 200,000 tons of granular polyethylene each year, compared with 100,000 in the rest of the USSR. The combine also produces 125,000 tons of propylene, 100,000 tons of benzene, and high-octane petrol from a hydro-treater. All its facilities were built by Linde of West Germany, except the polyethylene plant which was built by John Brown of the UK.

Another major combine is under construction at Zima in Eastern Siberia. An ethylene pipeline between Angarsk and Zima petrochemical plants was commissioned in 1981, and the first ethylene began to flow in December 1981. The Angarsk ethylene plant, using feedstock from the Angarsk oil refinery, has a rated capacity of 300,000 tons a year, but will not reach this level of output for several years. The pipeline runs for 200 kms to Zima, where the ethylene is converted into ethyl dichloride and then into vinyl chloride in a 270,000 tons a year plant built by the Uhde using a Goodrich/Hoechst process, and finally into PVC in a 250,000 tons a year plant built by Klockner using KHD/Pritchard engineering and a Huls process. These plants started up in 1982, and the Zima combine has been earmarked for expansion during the 1980s.

Once the technology has been mastered for the construction of these large technologically advanced plants, there is no reason why the output of chemical products should not grow extremely rapidly. Estimates based on the capacity of plant now under construction suggest that the 1985 plan targets for the production of 6 mn tons of plastics, 1.6 mn tons of chemical fibres and 150 mn tons of fertilisers are feasible.

As fuel is used primarily as a feedstock, it is unlikely that the rate of utilisation of fuel will fall significantly for chemicals as a whole. Fuel savings achieved through economies of scale and more advanced processes will be balanced by a larger share of fertilisers produced from natural gas, i.e. nitrogenous types. This can be deduced from the fact that the 1980 plan target for synthetic ammonia was 90 per cent greater than the actual output in 1975; the corresponding figure for fertilisers was 53 per cent.

This suggests that the share of natural gas in the total volume of fuel con- sumed by the sector will rise, and the greater use of casinghead gas, refinery gas and gas condensate will reduce the volume of straight-run distillation benzine.

It is possible to make a tentative estimate that fuel consumption by the chemicals sector will rise to 80 mn tsf by 1985, including 39 mn tsf of gas (33 bn cu m of gas), 38 mn tsf of oil-based feedstocks and 3 mn tsf of coal-based feedstocks.

THE HOUSEHOLD AND MUNICIPAL SECTOR

This section deals with the consumption of fuel by the population for the purposes of heating homes, cooking food, and providing hot water, and by municipal authorities for the heating of public buildings and cooking food in restaurants, etc.

In 1980, 1,500 heat-and-power plants and several thousand regional boilers and other centralised sources of heat served 70 per cent of all dwellings in the USSR, including 80 per cent in Russia and 60 per cent in the rest of the country, and provided 60 per cent of their heat requirements. The other 30 per cent of all dwellings were served by 120,000 house-boilers, group-heated boilers and domestic heating furnaces.

In 1965, the centralised sources of thermal energy (i.e. heat-and-power plants, industrial boilers, and large regional boilers with productivities of 20 G-cals and more a day) produced 853 mn G-cals of thermal energy, of which 114 mn (13.4 per cent) went to the household-municipal sector. In 1970, these figures rose to 1,296 mn and 212 mn (16.4 per cent), in 1975 to 1,767 and 327 mn (18.5 per cent), and the plan target for 1980 was 2,260 and 480 (21.2 per cent) mn G-cals. The target was underfulfilled; 2,260 mn G-cals were produced, of which the household and municipal sector used 460 mn. Thus the consumption of centrally produced thermal energy per person grew from 0.49 G-cals in 1965 to 0.87 in 1970, 1.28 in 1975 and 1.74 in 1980.

Consumption of thermal energy by the household-municipal sector from all sources grew by 25 per cent over the period 1970-1975 from 830 to 1,040 mn G-cals or from 3.4 G-cals per person to 4.1 G-cals. The share of this energy coming from centralised sources rose from 25.5 per cent to 34.1 per cent. It is natural that, with their heavy capital investment costs, thermal energy systems should be much more prevalent in the cities where they can be used more intensively, and in 1975 some 660 mn G-cals were consumed in towns (4.21 per person) and 380 mn G-cals in rural localities (3.84 per person). It is apparent that, as with the distribution of gas, the gap between urban and rural living standards is being rapidly narrowed.

The larger the city, the more economical the laying of a thermal energy system, and the larger the share of consumption that should be covered by heat-and-power plants. Thus Mosenergo covered 67 per cent of Moscow's household and municipal consumption in 1975. For the country as a whole, HPPs covered 27 per cent of heat requirements, and this was expected to rise to 33 per cent by 1980. In some new towns, created on the basis of a large industrial complex, HPPs covered the town's entire demand for steam and hot water.

Although the development of HPPs is more advanced in Moscow than in most other cities, even here much more needs to be done. The centralised supply of thermal energy is partly provided by peak-supply boilers, some of which have become base-suppliers, and operate for an unduly large number of hours. This is because the construction of HPPs is lagging behind and during the last the years plans to build HPPs were underfulfilled by 5,000 mw.

In other cities, the failure to make full use of the potential offered by HPPs stems from opposite reasons. In order to obtain the greatest possible savings of fuel, it is necessary to join up consumers to the stations as rapidly as possible by laying the thermal energy networks in good time. This, on the whole, has been accomplished, and the problem has been delays in the construction of industrial enterprises, which had led to HPPs largely on an uneconomical condensing cycle, primarily producing electricity at rates of utilisation of fuel higher than those of condensing power stations.

For the purpose of space-heating, the provision of central heating on a mass basis by means of thermal energy produced by HPPs, industrial and domestic boilers, are most economical.

The volume of kerosene sold to the population from State and Cooperative shops has been declining sharply from 2.7 mn tons in 1960 to 0.5 mn in 1980. Sales of coal rose steadily until 1971; since then they have grown very slowly to 30.2 mn tons in 1980. The consumption of coal has been rising at the expense of firewood, which was an important fuel for space heating, even in major cities, until quite recently. Sales of firewood to individual consumers in the countryside are still substantial, but no data are available.

Coal is used primarily for space heating and cooking. Few households use electricity for cooking; only 5 per cent of families possessed electric stoves in 1975. Gas is much cheaper, and with consumption of electricity at peak times often approaching crisis levels, the programme to connect households to gas mains as quickly as possible is receiving strong government support. By the end of 1981, 55.2 mn dwellings were receiving gas, compared with 41.7 mn in 1975 and 23.4 mn in 1970. While most gas for this purpose is despatched to large cities through large-diameter distribution pipelines, the production of liquified gas enables thousands of small localities to be supplied.

In 1980, 36.4 mn urban dwellings (80 per cent of the total) received gas compared with 28.8 mn in 1975. In oblast centres, normally large towns with multi-storey buildings, 80 to 85 per cent of flats are connected to the mains, and a further 5 to 10 per cent use liquified gas. In some cities like Moscow and Leningrad, practically all dwellings are served with mains gas. Natural gas has been used widely since 1965 when urban pipeline networks were begun. In large cities of 100,000 to 500,000 people, 70 to 75 per cent of flats use mains gas, and 5 to 10 per cent liquified gas. In small cities, 10 to 15 per cent of flats use mains gas, and 10 to 15 per cent liquified gas. The extent of the gas mains network is growing very slowly, mainly in new housing areas.

In 1980, 18.8 mn rural dwellings (70 per cent of the total) received gas compared with 12.8 mn in 1975 and 4.6 mn in 1970. Some 200 mn people now use gas, including 70 mn in the countryside.

The extent to which regions of the USSR are served with gas can be measured by calculating the number of gas-served dwellings per thousand of the population. It can be seen that dwellings in all the non-Russian regions of the USSR, with the exception of Central Asia (which is hampered by the high proportion of rural inhabitants in the total), are more likely to be served with gas than those of the Russian Federation. The Baltic states stand out as being particularly well served, followed by the Ukraine.

In 1980 there were 219 gas-served dwellings per thousand people in urban areas compared with 191 rural localities; the provision of gas to villages in Kazakhstan was particularly extensive.

Table 40

Gas Consumption for Residential and Municipal Purposes

Natural gas (bn cu m) households	11.5	16.2
Public buildings	21.7	23.5
Total	33.2	39.7
Liquified gas (mn tons) households	3.0	4.0
Public buildings	0.3	0.5
Total	3.3	4.5

The future rate of provision of gas to dwellings and public buildings is difficult to predict. The Eleventh FYP foresees continued growth in supply of natural gas delivered to flats and to municipal boilers, mainly in new mikro-rayons of large cities. This will accelerate the transfer from solid fuel to gas. In small and medium-sized towns connected to gas pipelines, it is necessary to speed up the construction of gas distribution networks, especially those serving four- to five-storey buildings. This will release liquified gas for single houses and rural areas. For towns where mains gas is not yet delivered, the decisions to connect them up to the system for delivery to the household and municipal sector must be taken after analysing the effectiveness of using gas in both industrial and domestic sectors. For rural areas, the greater use of liquified gas is planned.

Multi-storey flats, especially in new mikro-rayons, are highly effective and top priority customers of natural gas. However, the existing price system does not encourage the accelerated construction of urban distribution pipelines. At the present time, the prices of gas sold to households are subsidised, and in particular they do not include the costs of the construction and operation of a low-pressure distribution network with regulation points. Domestic consumers pay a price of 11 to 15 roubles per thousand cu m for gas. This is lower than the prices at which gas is sold to industrial enterprises, although heating boilers create seasonal irregularities in the consumption of gas which leads to higher costs, i.e. on the construction of underground storage reservoirs. This encourages the gas sales organisations to supply industrial enterprises more quickly, and dwellings more slowly. The "privilege prices" of gas to the domestic sector therefore lead to delays in supplying it with gas.

Raising the price level will lead to an acceleration of gas supply to the populace. At the moment, urban dwellings are equipped for food preparation with gas stoves using mains gas (40 per cent) or liquified gas (25 per cent), electric stoves (5 per cent) and coal-burning stoves (30 per cent). The users of mains gas spend, on average, two roubles a year. Raising the price of gas, even by 100 per cent, will have an insignificant effect on household expenditures, and will help to speed up the connection of dwellings to the gas mains.

While users of mains gas spend two roubles a year, those using bottled gas spend up to 10 roubles a year, the same as for electric stove owners.

Most large public buildings, such as hospitals, schools, shops, cinemas, etc., derive their heating from centrally produced thermal energy, and their direct consumption of fuel is very small. Restaurants account for most of the fuel used by public buildings, i.e. gas and heating oil for the cooking of food. Schools and hospitals also require fuel for this purpose.

In spite of the rapidly increasing supply of natural gas to the population, its overall consumption of fuel is growing very slowly, and consumption by municipal buildings is not growing much faster. This is because households and public buildings are covering their heating requirements to an increasing extent by electricity and centrally produced thermal energy.

The cost of fuel to households and public buildings rose by 61.2 per cent over the period 1966-1972. This was mainly due to the price rise of 1967, hence the very large increase of 87.7 per cent in the value of coal used in this way. Natural gas increased its share of total consumption from 10.1 per cent in 1966 to 17.2 per cent in 1972, when 582 mn roubles were spent on it. The use of oil products increased very slowly, and that of peat and shale nearly halved.

Table 41

Consumption of Fuel by Households and Public Buildings
(mn roubles at current prices)

	1966	1972	Increase
Coal	847	1,590	87.7
Oil	950	1,164	22.5
Gas	211	582	175.8
Peat & shale	92	51	-44.6
Total	2.100	3,387	61.2

% share of the total		
Coal	40.3	46.9
Oil	45.2	34.4
Gas	10.1	17.2
Peat & shale	4.4	1.5
Total	100.0	100.0

Sources: Treml and Kostinsky: Soviet Economic Prospects for the
          Seventies. The Soviet Economy in a New Perspective

The increase in the share of coal is because the price of coal rose faster than that of other fuels. If measured in terms of volume it is likely that coal's share would remain about the same.

The proportion of all fuel going to the sector, after rising to 11 per cent in 1966 from 9.4 per cent in 1959, fell to 8.6 per cent in 1972.

Table 42

Consumption of Fuel by Households and Public
Buildings as a Proportion of Total Consumption of Fuel

(%)	1959	1966	1972
Coal	10.7	12.6	12.7
Oil & oil products	6.9	9.4	5.3
Gas	32.3	12.8	14.5
Peat & shale	12.8	16.3	8.4
Total	9.4	11.0	8.6

Source: Calculated from Tremi and Kostinsky op. cit.

163

The most interesting feature of Table 42 is the substantial fall in the share of oil and oil products going to households and public buildings. This is largely due to the scrapping of small, inefficient boilers running on fuel oil and serving small areas of cities or blocks of flats. They have been replaced by the centralised production of thermal energy for central heating purposes.

The future consumption of fuel by the sector is influenced by many factors. Those likely to increase consumption include the increase in the population, and the replacement of coal by gas for cooking purposes as more and more dwellings are connected to the gas supply system. Those tending to slow down the increase in consumption include the increasing degree of urbanisation which enables expensive coal and oil to be replaced by cheaper natural gas, and the tendency towards the concentration of the production of thermal energy, firstly in centralised (as opposed to local) sources of thermal energy, and secondly in larger centralised producers. The new 1,350 mn HPPs in Moscow are considered optimal for large areas of concentrated housing and will eventually be repeated in Leningrad and other large cities.

In view of the large number of towns, not to mention villages, which as yet are not supplied with gas either from mains or in liquified form, it is reasonable to suppose that the number of dwellings provided with gas will continue to grow by 2 mn a year for the next few years. This assumption should be reinforced by the fact that some of the Soviet Union's largest cities like Omsk have still not been connected up to the pipeline system, and Novosibirsk (population 1.3 million) received gas by pipeline for the first time only in 1980.

If this rate of provision of gas is maintained, then 94 per cent of the country should be supplied by 1990. Efforts to extend the provisions of mains gas will be continued, and from 1985 onwards the consumption of liquified gas by private dwellings should start to fall. It can be estimated that by 1985, 63 mn dwellings will be using 19.3 bn cu m of natural gas and 5.5 mn tons of liquified gas.

If the low price of gas is maintained, it is unlikely that the use of electricity for cooking will continue to any significant extent. Electric power will be confined to lighting and the powering of domestic implements such as refrigerators, washing machines, televisions, etc.

By 1985, the use of firewood for cooking purposes should have practically disappeared, and the consumption of coal for this purpose will be declining. When it is remembered that, in terms of cost per calory of heat, coal is 7.25 times, kerosene eight times and firewood 17 times more expensive than gas, and that conversion to gas saves the average family 40 roubles a year, the urgency behind the gas supply programme can be understood.

Consumption of gas by public buildings should grow fairly slowly. They tend to be served by gas before dwellings because their large size means that greater savings can be made from the scrapping of oil and coal fired boilers. Consequently, most public buildings are already supplied with gas, and it is reasonable to suppose that consumption will grow at a rate of 3.5 per cent a year. On this basis, the household and municipal sector will require 47 bn cu m of natural gas and 5.2 mn tons of liquified gas in 1985.

By 1985, the campaign to build large heat-and-power plants will have raised the share of thermal energy from centralised sources to 61 per cent of the total volume of heat used by the household and municipal sector, i.e. 1,000 mn G-cals from 1,630. Domestic boilers of various types will account for the remaining 630 mn G-cals, compared with 685 mn G-cals in 1975.

There is little available information on the rate of consumption of fuel by domestic boilers in recent years, and it is probably pessimistic to assume that there has been no improvement in the 350 KgSF per G-cal registered in 1965. If it is assumed that the rate of consumption fell to 300 in 1975, and can be further improved to 275 by 1985, then the total volume of fuel required by these boilers will fall from 205 mn tsf in 1975 to 165 mn tsf in 1985. These estimates refer to the heating and hot water supply of dwellings and public buildings. It can be estimated that in 1980, fuel consumption for the heating of dwellings alone amounted to 150 mn tsf, although Izvestia puts it at only 120 mn tsf. Large volumes of fuel are also required for the provision of hot water, much of which is wasted in the Soviet Union because of its heavily subsidised price. Consequently, the average Soviet dwelling consumes 100-120 litres of hot water a day compared with 60 to 80 in Western countries.

Fuel consumption by the domestic and municipal sector has not fallen as rapidly as hoped by the planners. This is because delays in building HPPs have meant that, instead of closing small boilers down, the Russians have had to increase their number by 25,000 during the period 1971-1975, and this tendency continued into the 1976-1980 period. It is apparent that the USSR can make substantial fuel savings if it steps up the rate of HPP construction.

CHAPTER 12

Current and Forecast Foreign Trade in Hydrocarbons

It seems to be taken for granted by many Western observers that the volume of fuel exported by the USSR is regarded by Soviet planners as a residue after the country has satisfied its own needs. In fact, it is likely that exports play a major role in the planning balance, in financial rather than volume terms, and any shortfall in production makes itself felt through a reduction in fuel available for domestic needs rather than a decline in exports. Exports are needed to finance imports of equipment which the USSR could only produce itself at a prohibitive cost in terms of money and time, and are therefore considered more important than the domestic production of other, less vital, items.

It has been argued that the USSR is planning to increase its export of oil and gas while using atomic power and inefficient, but cheap, resources of coal to produce electricity. The additional cost to the economy would be more than covered by the benefits accruing from the imports of Western technology, particularly in labour-saving fields such as computers and automation. In addition, compensation agreements are being made with Western consortiums whereby large-diameter pipe and oil and gasfield equipment are obtained in exchange for long-term deliveries of oil and gas.

At the present time, crude oil dominates the export pattern of Soviet fuel, with exports of natural gas growing rapidly while those of oil products and solid fuels grow much more slowly.

Table 43

Exports of Fuel and Energy by the USSR

	1960	1970	1975	1978	1979	1980
Crude oil (mn tons)	17.8	66.8	93.1	121.1	117.7	115.8
Oil products (mn tons)	15.4	29.0	37.3	42.1	42.7	43.3
Coal and coke (mn tons)	14.9	28.7	30.3	30	28	26
Gas (bn cu m)	0.2	2.2	19.3	37	48.4	57.3
Total fuel (mn tsf)[1]	59.8	167.6	238.9	306.3	314.5	320.6
Electricity (bn kwh)	0.0	5.2	11.3	12.2	15.2	19

Sources: Statisticheskii Ezhegodnik Stran Chlenov S.E.V., various issues.

1 Narodnoe Hkhozyaistvo 1980, since 1976, authors estimates.

The principal feature of the table is the rapid growth in exports of oil and oil products from 33 to 159 mn tons over the period 1960-1980, or by 8.1 per cent a year. Exports of gas were due to treble over the period 1976-1980 to 58 bn cu m, or 13.3 per cent of production, and this target was almost achieved. Electricity is exported to Finland, Norway, Greece, Poland, Hungary and Bulgaria, and may eventually be sent to the GDR and West Germany.

The value of fuel exports has grown dramatically since 1975.

Table 44

Exports of Fuel by Value (mn roubles)

	Crude oil & products	Solid fuel	Natural gas	Electri- city	Total	Share of fuel in total export.
1973	2,405	432	92	100	3,027	19.2
1974	4,352	565	213	113	5,243	25.3
1975	5,908	1,005	451	159	7,523	31.4
1976	7,676	1,021	751	168	9,616	34.3
1977	9,400	1,046	1,023	194	11,663	35.1
1978	10,052	965	1,473	228	12,718	35.6
1979	14,517	983	2,089	280	17,869	42.2
1980	18,085	1,101	3,687	400	23,273	46.8

Source: Vneshnyaya Torgovlya

Fuel is exported to different regions of the world according to different motives. An important destination is the rest of Comecon, which took nearly half of Soviet petroleum exports in 1978, and where the principal motive is that of trade specialisation, although political considerations are also important. The USSR is the only Comecon member, apart from Rumania, to produce large quantities of oil, and it exports oil to these countries in exchange for industrial products. It has a political, as well as economic, obligation to supply the rest of Comecon, and accordingly it sells oil at prices lower than the world market price. According to the "Bucharest formula", prices amount to the average of the world market price for the previous five years, and this means that in 1980 Comecon countries paid only 60 per cent of the world price for oil bought under annual contracts from the USSR. During the first half of 1980, Poland, for example, paid 65 to 69 per cent of the world price. The aim of the Bucharest formula is to soften the impact of world price rises on the economies of Comecon countries. Oil prices rise regularly but gradually, their level for several years in advance can be roughly estimated by Comecon planners, and they are always significantly lower than world prices.

Means other than prices rises are needed to encourage the East European countries to economise on oil. These take the form of limits to the volume of oil sold under term agreements; the Eleventh FYP has stipulated that term sales will amount to 400 mn tons over 1981-1985. Thus, the level of term exports is not expected to grow beyond the 80 mn tons exported in 1980. But the USSR will supply additional oil at prices calculated from a formula based on Opec prices excluding those of the African producers. Such oil must be paid for in hard currency, but the Eastern European countries will still prefer Soviet to Opec oil because its supply is less subject to political upheavals, wars, etc. Soviet strategy, on the other hand, is to encourage them to look elsewhere for their oil, leaving more Soviet oil available for export to the West.

Prices of oil sold to Western countries have risen in line with world prices, and oil sales have grown to a level which enables the USSR to achieve a surplus on its trade with the OECD countries. Its surplus rose from $0.9 bn in 1979 to $3.2 bn in 1980, but fell back to $1.6 bn in 1981. In 1980, the Soviet Union sold 57 mn tons of crude and products, worth 9.11 bn roubles to the OECD, and 11 mn tons worth 1.0 bn roubles to developing countries.

Table 45

Destination of Petroleum Exports by Volume and Value

	1975	1976	1977	1978	1979	1980
Volume (mn tons)	130.4	148.5	161.1	163.1	160.4	159.1
Socialist countries	77.7	84.0	88	90	90	91
Capitalist countries	52.7	64.5	73	73	70	68
OECD	39.1	50.2	60.7	60.7	58.9	57.0
Others	13.6	14.3	14	12	11	11
Value (bn roubles)	5.9	7.7	9.4	10.1	14.5	18.1
Socialist countries	2.7	3.2	4.2	5.2	6.4	8.0
Capitalist countries	3.2	4.5	5.2	4.9	8.1	10.1
OECD	2.8	3.9	4.4	4.3	7.2	9.1
Others	0.4	0.6	0.8	0.6	0.9	1.0

Sources: Vneshnyaya Torgovlya, QECD Quarterly Oil Statistics, authors estimates.

In spite of forecasts by some Western observers that oil sales to the West are due to suffer a precipitous decline, they continued to grow until 1979, when they fell by 4 per cent compared with 1978, and by a further 3 per cent in 1980. Exports of crude oil to OECD countries fell from 36.3 mn tons in 1978 to 33.4 mn in 1980 and sales of products fell from 24.2 to 23.6 mn tons. This was due to a decline in the sale of diesel from 14.8 to 14.2 mn tons, and of fuel oil from 5.5 to 4 mn tons, while exports of naphtha rose from 2.6 to 3.8 mn tons.

The principal Western customers are Finland, France, West Germany and Italy, with the Netherlands and Belgium becoming more important in 1979 and 1980 as the Russians took advantage of surging spot prices on the Rotterdam and Antwerp markets.

Table 46   Oil Sales to OECD Countries (crude and products)

Volume (mn tons):	1977	1978	1979	1980	Value ($ mn) 1977	1978	1979	1980
Finland	10.66	9.76	10.45	9.70	1,041	983	1,677	2,438
France	4.76	4.75	6.12	8.38	472	620	1,468	2,170
Netherlands	2.30	2.45	5.66	7.24	237	287	1,183	1,094
Italy	9.68	8.95	6.38	6.90	887	1,067	1,135	1,704
West Germany	8.04	8.89	9.33	6.88	833	1,098	2,053	1,006
Belgium	1.60	3.52	2.72	4.26	159	219	375	574
Switzerland	2.55	3.09	2.21	2.63	314	201	360	826
Sweden	3.55	4.32	4.30	2.35	340	344	696	557
Austria	1.93	1.98	1.73	1.51	190	189	269	386
Spain	0.64	0.45	0.88	1.38	53	123	126	353
Denmark	2.14	2.54	2.05	1.30	212	230	401	342
Greece	2.14	2.10	1.08	1.27	85	234	186	69
United Kingdom	4.66	4.49	2.89	1.21	480	449	477	408
Japan	0.59	0.56	0.65	0.53	76	107	146	164
Portugal	1.00	0.80	0.83	0.52	104	75	118	130
Iceland	0.43	0.38	0.37	0.36	49	na	na	89
Norway	0.83	0.83	0.82	0.20	85	68	101	48
Turkey				0.19				79
Ireland	0.40	0.41	0.37	0.18	37	48	80	38
USA	0.69	0.44	0.06	0.03	64	150	244	17
Total OECD	58.59	60.71	58.91	57.00	5,739	6,492	11,095	12,499

Sources: Volume: OECD Quarterly Oil Statistics
          Value: OECD Statistics of Foreign Trade Series C.

For some countries, the figures for volume and value obviously do not match, and the value of imports is probably understated.

In fact, the volume of Soviet oil reaching the West is greater than shown in Table 46. During the last few years, the Russians have been pushing products rather than crude to the West, and their sales of products via Eastern Europe have been increasing. They have adopted this policy partly because their own refineries are operated close to capacity, and partly because it is cheaper to deliver crude by pipeline to Schwedt refinery in East Germany and to Czech refineries where it is refined for re-export. An important factor assisting this trade has been the relatively successful efforts by Eastern European countries to reduce their own crude requirements by upgrading their refineries to produce their light and middle product needs. Consequently, 12.5 mn tons of Soviet oil, including 1.4 mn tons of crude and 11.1 mn tons of products found its way to the West in 1980 from Eastern Europe, and this compares with 8.4 mn tons in 1979 and 6.7 mn tons in 1978. Accordingly, the total volume of Soviet oil consumed by OECD countries rose from 67.4 mn tons in 1978 to 69.5 mn in 1980.

In 1981, however, there were persistent rumors that the Russians were planning to revert to direct sales to the West by cutting their deliveries to Eastern Europe by 10 per cent. This policy, connected with the recent commissioning of new capacity at refineries in the western part of the country like Lisichansk and Mazheikiai will also have the purpose of encouraging Eastern Europe to economise still further, or to take advantage of the falling prices of crude on the world market.

The Soviets have been badly affected by the slump in world crude prices. Their term prices to the West have remaind ahead of OPEC market prices, reflecting the higher quality of Soviet crude, and have kept fairly close to the price of North Sea Forties oil.

Table 47

Changes in the Price of Romashkino 32.4° Crude Delivered to Finland, $ a barrel

	Romashkino 32.4°	Saudi Light 34°	UK Forties 36.5°
1980 May	34	26	3$.25
June	34	28	36.25
July	36	28	36.25
August	36	28	36.25
September	36	28	36.25
October	36	30	36.25
November	36	30	36.25
December	36	32	36.25
1981 January	36	32	36.25
February	36	32	36.25
March	39.25	32	39.25
April	39.25	32	39.25
May	39.25	32	39.25
June	39.25	32	39.25
July	38.58	32	35
August	36.78	32	35
September	36.78	32	35
October	35.80	32	35
November	35.49	32	35
December	35.49	34	36.50
1982 January	35.49	34	36.50
February	35.49	34	36.50

Source: Oil and Gas Journal.

Exports of gas are rising rapidly, and the Russians have been pushing their gas sales with a view to increasing their hydrocarbon exports now that oil exports are beginning to fall.

Soviet gas exports flow across Czechoslovakia in the so-called Transit Pipeline which carried 37 bn cu m in 1980 to Western and Eastern Europe. In 1981, its third string was completed, raising its capacity to 53 bn cu m a year. Exports to Western Europe then travel through a pipeline system originating at Baumgarten on the Austrian-Czech border. Built by Ruhrgas, OMV of Austria and Gaz de France, it was originally intended for the Consortium Project, under which Iran would sell casinghead gas to the USSR, with the Russians selling a similar amount to Western Europe.

The construction of another string of the Transit Pipeline is to begin in 1982. It will be part of the Yamal pipeline from Urengoi to Western Europe, and will carry 40 bn cu m of gas a year. Under the terms of the contracts concluded so far with West Germany (for 11.2 bn cu m a year, including 0.7 bn for West Berlin), Italy (8.5 bn cu m) and France (8 bn), the Soviets can expect to receive $7.2 bn a year at 1981 prices when the pipeline is operating at full capacity. Over the 25 year period of the contracts, they should receive a revenue of $180 bn at 1981 prices, which may prove sufficient to pay off Comecons debts to the West.

Table 48

Export of Natural Gas

	Volume (bn cu m):				Value (mn $)			
	1977	1978	1979	1980	1977	1978	1979	1980
Socialist	19.4	20.7	28.5	31.7	818	1,014	1,717	2,803
Csechoslovakia	6.2	6.7	8.6	8.7	259	332	513	683
East Germany	4.7	4.8	5.6	6.5	200	235	335	545
Poland	3.5	3.5	5.1	5.3	146	170	302	470
Bulgaria	3.7	3.9	4.0	4.5	155	189	239	352
Hungary	1.3	1.4	3.3	3.8	58	67	194	347
Yugoslavia	-	0.4	0.9	1.9	-	20	75	264
Romania	-	0.0	1.0	1.0	-	1	59	142
Capitalist	12.0	16.3	19.9	25.6	593	1,133	1,480	2,852
West Germany	4.0	6.9	8.7	11.8	197	465	585	1,115
Italy	3.6	3.7	5.3	7.1	178	245	392	784
Austria	3.0	2.7	2.8	3.0	150	212	259	331
France	0.3	2.1	2.1	2.8	13	146	174	456
Finland	1.1	0.9	0.9	0.9	54	65	70	146
Total gas exports	31.4	37.0	48.4	57.3	1,411	2,147	3,198	5,645

Sources: Calculated from Vneshnyaya Torgovlya, and author's estimates

Oil and Gas Exports: Forecasts to 1985 and 1990

In 1980, domestic consumption of oil in the USSR amounted to an estimated 444 mn tons, or 4.2 per cent more than in 1979. It is likely that the average rate of growth of consumption will be no more than 2 per cent a year, and oil consumption will rise to 490 mn tons in 1985 and 540 mn in 1990.

This forecast has been reached after an examination of the likely development of the most important fuel-using sectors, and it assumes that their production targets for 1985 will be largely fulfilled. It also assumes that the major policy change of 1980 to concentrate on gas production will be followed through. Some oil-burning power stations have already been fully converted to gas, and others are being adapted so as to burn gas in summer. The current five-year plan calls for the scrapping of an unprecedented volume of thermal power station capacity with new sets burning coal (at Ryazan, Zuevka, Ekibastuz, Perm etc.) and gas (Azerbaidzhan, Surgut, Pechora and most heat and power plants) replacing obsolete sets burning oil. While the process of converting power stations to gas is simple and quick, the replacement or conversion of a very large number of much smaller oil-burning facilities will prove more difficult and time-consuming. Nevertheless, it is hoped to make substantial savings at power stations alone; as Table 34 shows, their consumption of fuel oil is planned to decline by nearly 40 mn tons of standard fuel between 1980 and 1985. Needless to say, the successful accomplishment of the atomic power programme is absolutely crucial to any attempt to forecast the future demand for fuel oil. If the atomic stations are not built on time, then the obsolete sets will be kept running, and oil consumption will not fall as much as planned.

If the rather demanding plans for the installation of secondary capacity at the oil refineries can be achieved, then it should be possible for the annual production of petrol and diesel to grow by 5.5 per cent a year from 188 mn tons to 245 mn tons over 1981-1985. Even with significantly higher exports of these products in 1985, an average annual growth rate for domestic consumption of 5 per cent is possible. In addition, the consumption of naphtha should rise very sharply.

If the current FYP targets are basically fulfilled, then the USSR should be in a position to export 140 mn tons in 1985 compared with 159 mn tons in 1980. It will hope to limit its contracted exports of crude and products to Eastern Europe to 80 mn tons a year over 1981-1985, although it will undoubtedly bail out any countries in economic difficulties by making additional volumes available. During periods of low world prices, it will encourage its Eastern European allies to negotiate barter deals with other countries for oil, but in any case it can be estimated that by 1990, about half of Eastern Europe's oil will be coming from outside Comecon. There are indications that these countries are not happy with the prospect, preferring the security of supply and the lower price of Soviet oil. Even Romania has begun purchasing Soviet oil, with imports rising from 0.4 mn tons in 1979 to 1 mn in 1980 and 1.5 mn in 1981. It is possible that Soviet oil exports to Eastern Europe may rise during the 1981-1985 period, but this is not likely given the success in reducing oil consumption rates that some of the countries have obtained, and the probability of sharply higher imports of natural gas and electricity from the Soviet Union.

Other Communist countries will receive 10 to 15 mn tons a year. Most of it will go to Yugoslavia and Cuba, with the rest receiving small amounts of products.

Exports of oil to the West can be expected to fall slightly to 45-55 mn tons a year by 1985, and should fall sharply thereafter to as little as 30 mn tons in 1990. By 1985, the Yamal pipeline will be operating at capacity, and gas sales to the West should amount to 70 bn cu m a year. The Russians are hoping to export 100 bn cu m in 1985 and 180 bn in 1990. They will push their gas exports to Eastern Europe, thereby hoping to save oil for the more lucrative Western markets.

This author rejects the view that the Soviet Union is running out of oil, or that it does not have (and cannot develop) the technology to exploit its reserves. It will remain an important supplier of hydrocarbon fuels to the West for well into the 21st century. It is by far the world's largest oil producer, and will shortly become its largest gas producer, and as such has a powerful influence on world markets. Whether that influence is beneficial or detrimental to Western interests depends primarily on the West. The Yamal pipeline project is a prime example of how the Soviet Union's vast potential can be harnessed to the mutual benefit of East and West.

Bibliography

This report has been written on the basis of information obtained from the following:

Periodicals

Pravda (daily)
Izvestia (daily)
Sotsialisticheskaya Industriya (daily)
Trud (daily)
Vneshnyaya Torgoviya (monthly, with an annual yearbook)
Ekonomicheskaya Gazeta (weekly)
Neftyanoe Khozyaistvo (monthly)
Ekonomika Neftyanoi Promyshlennosti (monthly)
Gazovaya Promyshlennost (monthly)
Zheleznodorozhnaya Transport (monthly)
Statisticheskii Ezhegodnik Stran Chlenov S.E.V. (annual)
Khimicheskya Promyshlennost (monthly)
OECD Quarterly Oil Statistics
OECD Bulletin of Trade Statistics
Narodnoe Khozyaistvo SSSR (annual)
Narodnoe Khozyaistvo RSFSR (annual)
Narodnoe Khozyaistvo Azerbaidzhanskoi SSR (annual)
Narodnoe Khozyaistvo Kazakhstoi SSR (annual)
Narodnoe Khozyaistvo Turkmenskoi SSR (annual)
Planovoe Khozyaistvo (monthly)
Voprosy Ekonomiki (monthly)
EKO (Ekonomika i organizatsiya promyshlennogo prouzvodstva) (monthly)
Stroitel'stvo truboprovodov (monthly)
BBC: Summary of World Broadcasts, Weekly Economic Report - Part 1 (weekly)
Oil and Gas Journal (weekly)
World Oil (monthly)
Platts Oilgram News (weekly)
Soviet Geography (monthly)

Other Publications

Muravlenko & Kremneva, "Sibirskaya Neft," Moscow 1977
Yegorov, "Ekonomika neftepererabotyvayushchei i neftekhimicheskoi promyshlennosti",
  Moscow 1974
Malyshev, "Ekonomika organizatsiya i planirovanie neftepererabotyvayushchikh
  zavodov", Moscow 1975
Feygin, "Neftyanye resursy, metodikha ikh issledovaniya i otsenki", Moscow 1974
"Soviet Economic Prospects in the Seventies", a compendium of papers submitted
  to the Joint Economic Committee of the US Congress, Washington 1973
Gorshkov, "Tekniko-ekonomicheskie pokazateli teplovykh elektrostantsii," Moscow
  1974
Savelev, "elektroenergeticheskaya baza ekonomicheskikh rainov USSR", Moscow 1974
Bobrovskii, "Narodnokhozyaistvennaya effektivnost' khimizatsii proizvodstva",
  Moscow 1968

Bushuev "Khimicheskaya industriya v svete reshenii XXIV sezda KPSS", Moscow 1976

Yegorov, "Ekonomika neftepererabotyvayushchei i neftekhimicheskoi promyshiennosti" Moscow 1974

Bulyanov, "Neft i gaz v narodnom khozyaistve SSSR", Moscow 1977

Mazover & Probst, "Razvitie i razmeshchenie toplivnoi promyshiennosti", Moscow 1975

Nekrasov & Pavienko, "Energetika SSSR v 1970-1975 godakh", Moscow 1976

The Market for LPG
in the 1980s

INTRODUCTION

In a world which seems likely to run short of oil in the current decade it is inconceivable that there should be a surplus of liquefied petroleum gas. That, nevertheless, was the conclusion of a study which the EIU published in early 1977, examining the situation up to 1985, and it is a conclusion which - for the period until then, at any rate - is now the accepted wisdom.

The present study is not intended as an updated version of the earlier one, although it covers some of the same ground. It is an attempt to examine the market ten years hence in the context of the opening sentence above: in other words, to consider in what geographical areas and end uses LPG is likely to be absorbed if it is assumed that its price relationship with other fuels/feedstocks will move in a manner permitting absorption. Whilst it is indeed probable that a surplus will exist in the mid 1980s, despite some price adjustments, some rolling in of higher import with lower domestic prices, the extent to which it will be manageable will largely depend on how potential users foresee the longer term situation: they are not going to invest in the special facilities required for the transport and storage of LPG if the economics of the investment seem likely to become adverse by 1990. But if no new investment is made, the surplus will persist to 1990.

The implications are twofold. From the user's point of view investment should have an inbuilt flexibility, i e not be dedicated solely to the use of LPG - a condition which largely defines where it is likely to be made. But flexibility has a cost. From the producer's point of view, some long term commitment on prices is, for that reason, required, and also because intermediate between him and the user is the provider of transport, reception terminals and storage - to the extent, at least, that the latter two are not the same. The disposal of large volumes of LPG approximates to the closed loop system of trading that applies in the case of LNG, and some price certainty is an essential element in that system.

This conclusion may seem a truism.[1] In its defence may be quoted the following statement by an Opec representative: "It is the responsibility of the consumers to make LPG economically attractive for the producers". It has, moreover, become something of a shibboleth among Opec members that LPG is by way of being a "noble" fuel which deserves a noble price, while more than one article has been published pointing out - with the implication that it is unnatural or unfair - that in terms of calorific value LPG has fallen in price in recent years in relation to crude oil[2]. A fuel or feedstock is, however. only noble to the extent that it is rare, a quality which will ensure that its use is confined to markets in which it can command a premium price. If it has to seek a wider market, its nobility is impaired and its price must fall to a level that will clear the market. That is the situation with LPG.

---

1 "Opec Bulletin", August 1980, page 25. 2 Eg ditto. September 1980, page 24.

CHAPTER 1

Western Europe

Supply of LPG: the historical situation

Traditionally almost all the LPG used in Europe has been produced in refineries.
Some of the iso-butanes are separated from the normal butanes for dehydrogenation, to be used in the production of synthetic rubber, and as much of the remainder
as possible is blended into gasoline up to the limit of vapour pressure specifications.
The rest of the butanes and the propane have invariably been regarded as by-products.

Some of it is burnt as refinery fuel, depending on the time of year and the refinery's
location vis-à-vis market outlets - the latter tending to be circumscribed by the
high cost of storing and transporting LPG compared with liquid fuels. High-purity
LPG has a special use in metallurgical industries, and in certain others where
clean or direct heat application is important. but the major part of production has
been sold as a domestic fuel (chiefly in bottled form and in areas where natural
gas is not available) and for agricultural and general industrial use. Generally
speaking, LPG's premium qualities have not been recognised, in terms of price,
in European fuel markets. Since in many uses consumption is seasonal, summer
surpluses have to be stored, burnt or exported, and winter deficits met by
supplementary imports.

In 1973 Western Europe[1] was a small net exporter of LPG - to the tune of about
0.3 mn tons. Following the oil price rises at the end of that year, which were
not fully reflected in the price of LPG vis-à-vis those of competitive fuels, the
area became a net importer: 0.6 mn tons in 1975. when economic activity was at
a low level and consumption of LPG, as of oil products generally, fell below the
1973 level. Net imports are estimated to have risen to nearly 2 mn tons by 1979.

The first table below shows the trends in supply. As the second table indicates
(even allowing for the discrepancy pointed out in the footnote), much of the import/
export trade is intra-area: coastal traffic in small tankers. Nevertheless, identified
imports from Opec members reached 1.35 mn tons in 1979, almost 10 times the
1973 supply from that source.

---

1 It should be noted that throughout this section Western Europe excludes Turkey
and Iceland. for which no data are available on a basis comparable with those for
other OECD members.

Western Europe: Supply and Disposal of LPG
('000 tons)

	1975	1977	1978	1979
Refinery production	(11,830)	13,476	13,560	14,360
Refinery consumption	(-200)	-232	-250	-286
Net refinery output	11,630	13,244	13,310	14,074
Gas plants & other sources	737	894	(1,060)	(1,120)
Imports	2,617	3.156	3,576	4,267
Exports	-2,062	-2,553	-2,335	-2,278
Consumption[a]	12,922	14,741	15,611	17,183

a Calculated from the data in the table, i e excluding stock changes
and statistical errors in individual items. ( ) Estimated by the EIU.

Source: OECD.

Western Europe: Sources of LPG Imports
('000 tons)

	1975	1977	1979
Intra-OECD Europe	1,259	2,182	2,700
Eastern Europe	231	359	238
North America	9	26	48
Venezuela	53	115	152
Other W Hemisphere	4	7	62
Algeria	62	118	294
Libya	7	110	129
Kuwait	4	28	120
Saudi Arabia	17	133	548
Other Middle East	1	16	15
Indonesia	–	7	98
Other or unidentified	765	45	92
Apparent statistical error	+205	+10	-229
Total	2,617	3,156	4,267

NB: Figures for 1975 and 1977 are derived from OECD,
as is the 1979 total. Details for 1979 are collated from
the trade returns of member countries. The "apparent
statistical error" represents the difference between
members' figures and those for OECD as a whole, as
shown in the previous table. It is clear that the figure
derived from the trade returns of 2.7 mn tons of intra-
OECD European imports in 1979 is incompatible with
OECD's figure for total exports in that year of under 2.3
mn tons. It has not been possible to resolve this discrepancy.

Future supply

Except for Turkey (which is not included in this section) and a doubtful project in
Italy, there are currently no plans for additional distillation capacity in Western
Europe. In other words, any increase in refinery output in the immediate future
will result from an increase in the utilisation of existing capacity, above the 1979
level of 60 per cent. On the other hand, some 15 mn tons/year of catalytic

CHART 1.

Western Europe: LPG Imports

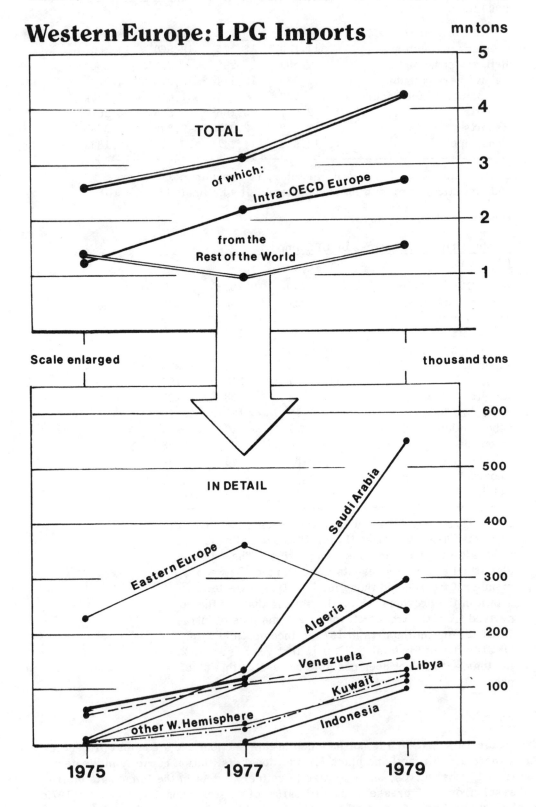

mn tons

TOTAL

of which:

Intra-OECD Europe

from the
Rest of the World

5

4

3

2

1

Scale enlarged

thousand tons

IN DETAIL

Saudi Arabia

Eastern Europe

Algeria

Venezuela

Libya

Kuwait

other W. Hemisphere

Indonesia

600

500

400

300

200

100

1975 1977 1979

cracking capacity was under construction or planned at end 1979, on top of the 45 mn tons already installed. Assuming 85 per cent utilisation of this new capacity, output of LPG would increase by approximately 1.8 mn tons gross. In 1985-90 an increase in cracking capacity, and of LPG yield from that source, of at least a similar order of magnitude is probable. To be set against this increase is the requirement for butane for blending into the additional gasoline produced by the new catalytic cracking facilities. In the table below the addition is shown net of this requirement.

A further consideration is increased refinery use of butane to produce alkylate as a high-octane blending component. According to one estimate, based on known alkylation projects, the quantity required may reach 2.7 mn tons by 1985, approximately 1 mn above the 1979 level. Beyond 1985 there is the question of the possible development of methyl tertiary butyl ether (MTBE) and tertiary butyl alcohol (TBA) as high-octane blending components alternative to alkylate, a question complicated by the value at that time of butane for petrochemical use vis-à-vis the production of alkylate: more alkylation capacity is unlikely to be installed if the value of butane to the refinery seems likely to exceed fuel oil equivalent. It is here assumed that butane for the production of alkylate will by 1990 be in the range 1-2 mn tons above the 1979 level.

First, however, there is the question of the level of crude throughput in 1985 and 1990. On the basis of submissions made by member states to the IEA - and with an allowance for consumption in France (the higher estimate in its latest energy plan) and in the other non-IEA members, Portugal and Finland - oil consumption in Western Europe is projected to increase at an average annual rate of 1.7 per cent a year in 1978-85, falling to 1.3 per cent in 1985-90. These rates are applied in the table below in arriving at refinery throughput. In respect of 1990, however, there is also the possibility that any increase in consumption above the 1985 level may be met by export refineries in the Middle East and North Africa; a range is therefore shown for that year.

Western Europe: Refinery Yield of LPG
(mn tons)

	1979	1985	1990
Refinery throughput[a]	694.8	735.0	735-755
LPG output on 1979 basis	14.4	15.2	15.2-15.6
Net LPG output from new catalytic crackers[b]	-	1.0	2.4-2.2
Total	14.4	16.2	17.6-17.8
Less refinery use of fuel[c]	0.3	0.3	0.3-0.4
Less additional butane for alkylation[b]	-	1.0	1.0-2.0
Net refinery supply	14.1	14.9	15.2-16.5

a Increase of 1.7 per cent a year to 1985 from the 1978 base level of 655.6 mn tons. b On assumptions set out in text. Yield from catalytic crackers in 1979 is included in the 14.4 mn tons of output shown, which is also net of butane used in alkylation units. c On proportional 1979 basis.

In addition to LPG from refineries, product is also extracted from natural and associated gas in France, Italy, Spain (imported Libyan LNG), and the UK. The net quantity available from these sources amounted to under 0.9 mn tons in 1977, the latest year for which OECD data are available. There has been little change since then except in the UK, where, with increased oil production in the North Sea and use of gas processing facilities, the quantity extracted rose from 208,000 tons in 1977 to 466,000 in 1979.

Gas separation and fractionation installations at Flotta, Sullom Voe and St Fergus/Mossmoran are expected to make approximately 2 mn tons/year of LPG available by 1985, a level that will probably fall slightly by 1990. In addition, the gas gathering system proposed by the British government for fields not served by the existing and planned crude oil and associated gas pipelines to these three terminals is forecast as yielding an additional 1.7-2.2 mn tons (from the UK sector only: the system might conceivably take in product from the Norwegian sector as well). In theory this system will be in place by 1984, but even on that, perhaps optimistic, basis, it is unlikely to reach full capacity in 1985. It is here assumed that around 2-3 mn tons of LPG will be available from the entire UK portion of the North Sea by 1985, and 3-4 mn by 1990.

In so far as the Norwegian sector is concerned, all the LPG and some of the ethane from the Ekofisk group of fields are separated at the production platforms and injected into the crude oil line to Teesport in the UK. where installations exist for their fractionation. After allowance has been made for the ethane, propane and n-butane shipped back to Norway to provide feedstock for the olefin cracker at Rafnes, it is estimated that in the early 1980s some 0.2 mn tons/year of n-butane will be exported. This export will be gradually reduced by the mid 1980s. By that time, however, an additional 0.1-0.3 mn tons/year of LPG may be available along with the crude oil from the Valhall field, which would be transported in the Ekofisk-Teesport pipeline. It is here assumed that export availability from the Norwegian sector will not exceed 0.3 mn tons in 1985.

By 1990 facilities will have been installed to exploit NGLs in the Statfjord field, the recent Shell discovery in Block 31/12 (where gas composition has not been announced) and other discoveries farther south. It has not yet been decided where the terminal for this Norwegian trunkline system will be located. We quote Phillips Petroleum on the question of NGL availability from the system:

"A substantial NGL production, which at yearly peak may reach one or more millions of tons, should not be impossible. Regardless of whether these NGLs become available in Emden, on the Shetland Islands or elsewhere, it can be expected that some of this production will be used as a feedstock basis for petrochemical production in Norway."

As is noted in a later section, Norsk Hydro and Saga Petroleum have plans for olefin crackers on the west coast of Norway - in connection with which considerable lobbying is under way to have the pipeline cross the Norwegian trench and terminate at Sotra, south-west of Bergen, or Mongstad. Some of the available NGLs will also be required at Rafnes, to replace a falling supply from Ekofisk-Valhall or supplement it, if the cracker is enlarged. On the whole, therefore, exports of LPG from the Norwegian sector as a whole appear unlikely to be more than 0.5-1 mn tons by 1990.

182

In summary then, the availability of LPG in Western Europe seems likely to be as follows.

Western Europe: LPG Availability
(mn tons)

	1979	1985	1990
Net refinery supply	14.1	14.9	15.2-16.5
Gas plants[a]	1.1	2.9-3.9	4.0-5.5
of which:			
UK North Sea	0.5	2.0-3.0	3.0-4.0
Norway[b]	-	0.3	0.5-1.0
Total	15.2	17.8-18.8	19.2-22.0

a Partly estimated. b Export availability.

Western Europe: consumption of LPG

Consumption of LPG in OECD Europe (excluding Turkey and Iceland) had by 1977 regained its 1973 level, and has since then risen strongly, by 9.3 per cent in 1978 and by 10.8 per cent in 1979. The average annual increase over the six year period 1973-79 was 3.3 per cent compared with 8.8 per cent in 1967-73. The table below shows comparative growth rates by end-uses in the two periods.

Western Europe: Growth of LPG Consumption by End-Uses
('000 tons)

	1967	1973	Average annual % growth 1967-73	1979	Average annual % growth 1973-79
Gas making	1,504	995	-	493	-
Road transport	313	659	13.4	1,375	13.1
Chemical (petrochemical)	1,000	2,161	13.6	3,388	7.8
Other industry	1,310	3,335	16.9	4,048	3.3
Commercial/domestic	4,173	6,634	8.0	7,252	1.5
Agriculture & other	110	156	6.0	360	14.9
Total	8,410	13,940	8.8	16,916	3.3

Fuller information, by country and end-uses, of consumption in 1967-79 is appended to this report. Data on 1978 and 1979 have been compiled from member country sources, and have had in some case to be estimated; they may not always be comparable with OECD data for earlier years but are unlikely to distort the overall picture.

Traditional uses.   In industry other than chemicals/petrochemicals, the commercial/ domestic sector, agriculture,etc,consumption of LPG in Western Europe rose from 10.13 mn tons in 1973 to 11.66 mn tons in 1979, at an average annual growth rate of 2.4 per cent. This rate was associated with real GDP growth averaging 2.6 per cent. GDP seems unlikely to rise at more than 2 per cent in 1979-85 but may, on optimistic assumptions, reach 2.5 per cent in 1985-90. If LPG consumption in traditional uses maintains its 1973-79 relationship with GDP, it would imply an increase in consumption from 11.66 mn tons in 1979 to 13 mn tons in 1985 and 14.9 mn tons in 1990.

CHART 2.

LPG : Western Europe

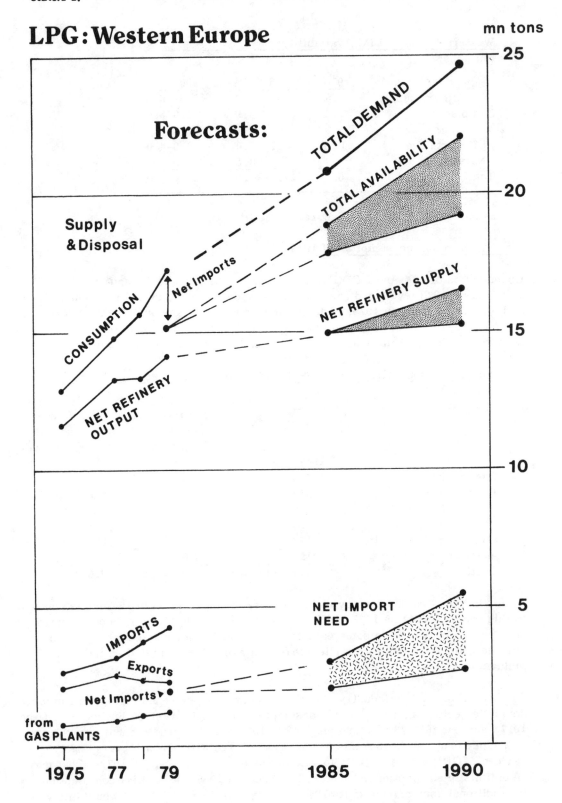

mn tons

Forecasts:

TOTAL DEMAND

TOTAL AVAILABILITY

Supply & Disposal

Net Imports

CONSUMPTION

NET REFINERY SUPPLY

NET REFINERY OUTPUT

IMPORTS

Exports

Net Imports ▶

NET IMPORT NEED

from GAS PLANTS

1975 77 79 1985 1990

25
20
15
10
5

Motor fuel. LPG has for a long time been used in Italy as a fuel for internal combustion engines, but it is in the Netherlands that consumption has increased most rapidly, from 96,000 tons in 1973 to 446,000 in 1979. Its wider use in Europe as a whole is difficult to forecast, given that its attraction stems from the lower level of tax currently levied on it: inevitably, such favourable treatment would rapidly come to an end if government revenues from gasoline were affected. Thus, for example, when in late 1978 the French government decided to authorise the use of LPG as a road fuel, it introduced regulations to restrict its use to vehicle fleets: vehicles had to be licensed, dual fuel systems were ruled out, and stations for LPG supply had to obtain special authorisation. At the end of 1979 it was estimated that only about 1,000 vehicles in France were operating on LPG, though the number was expected to rise steeply in 1980. Even so, the government then revised downwards its forecast of a year earlier, to an expected consumption of 200,000-250,000 tons/year by 1983, and 350,000 tons by 1985. Essentially the measure was considered as a temporary one, related to an excess of LPG availability in the local market as Spain, the former major outlet, became more self sufficient; in fact, the idea that LPG might be imported for the purpose seems to have been eschewed by the French government.

Meanwhile in West Germany some of the major oil companies have begun to install LPG pumps in certain service stations. The latest is Deutsche BP, which in autumn last year equipped 20 of its service stations in major cities with the requisite installations. According to the company, sales of LPG as a motor fuel in the Federal Republic are likely to rise to 1.5 mn tons/year in the next 5-7 years and account for about 6 per cent of motor fuel consumption. For Western Europe as a whole, Shell has forecast that LPG consumption may rise at an average annual rate of 6.5-7.1 per cent a year, compared with a growth in the motor fuels market as a whole of 1.2-1.9 per cent, and account for 4 per cent of that market by the year 2000. In fact, such a growth rate would be well below the rate of increase in 1973-79, which was of the order of 13.1 per cent a year. In that period, however, motor gasoline consumption was rising at an average annual rate of 2.6 per cent, and there must be considerable doubt, which Shell obviously shares, about such a high growth rate obtaining in the future.

On the assumption that gasoline consumption rises by 2 per cent a year in 1979-90, i e keeping pace with the postulated rise in GDP in 1979-85 but falling behind it thereafter, and that LPG use rises in a relationship with gasoline similar to the 1973-79 experience, demand for LPG as a motor fuel would amount to nearly 2.5 mn tons by 1985 and 4 mn tons by 1990. This forecast is, however, contingent upon governments continuing to tax LPG at a lower rate than gasoline. In that connection it is worthwhile examining two current situations, in the UK and West Germany.

In the UK (and also in Sweden and Australia) the tax on LPG as motor fuel is half that on gasoline on a gallonage basis. Without that concession there would in fact be little possibility of LPG's competing. A simple calculation will illustrate the point. In this illustration, Case I represents the situation where LPG is taxed at the same level as gasoline on a calorific basis. Case II represents the range of actual situations at those few UK service stations where LPG is available.

(p/imp gal)	Gasoline	Case I LPG	Case II LPG
Retail price	130.0	100.0[a]	90.0-103.0
Less VAT	17.0	13.0	11.7-13.4
Less tax	45.5	35.0	22.75
Wholesale price	67.5	52.0	55.55-66.85
Less retailer's margin	6.5	10.0	10.0-20.0
Distributor's price	62.0	42.0	45.55-46.85
= p/1,000 kcal	1.64	1.44	1.56-1.61

a Based on calorific equivalence to the motorist.

Margins range in the case of gasoline from 5.5 to 7.5 p/gal, and in the case of LPG from 10 to 20 p. In the former case the mean has been used in the table above, although in fact the average is nearer the bottom than the upper end of the scale. In Case I the low of the range has been used, presenting the most favourable case for LPG. As Case II shows, the higher price at the pump is associated with higher storage costs, i e a higher price has to be charged when throughput is low and unit storage costs high as a result.

In terms of calorific equivalence the distributor's price for LPG in the theoretical Case I is 1.44p/1,000 kcal. That compares with a distributor's price for gas oil concurrent with the above prices (third quarter 1980, excluding tax) of 1.43 p/1,000 kcal to industry and somewhat higher to domestic users. The actual price obtained for LPG by the distributor was around \$450/ton (at £1 = \$2.40).

In West Germany, LPG is taxed at the service station at approximate calorific equivalence with gasoline, as in Case I above: 33 pf/litre compared with 44 pf for gasoline, while the price at the pump, of 89 pf/litre, equates in calorific terms with the low of the price range for self service regular gasoline of around 115.5 pf. As a result of this higher tax (in relation, that is, to the UK), plus VAT of 10 pf and a retailer's margin of 5 pf, the distributor's receipt is only 41 pf/litre, 1.32 pence per 1,000 kcal or \$374/ton at the current exchange rate between the dollar and the pound. It is, however, above the gas oil parity of 1.27 pence per 1,000 kcal.

It may be said in conclusion that whereas (at the prices shown) the UK distributor would probably be able to import LPG for use as a motor fuel from the Middle East at its current price (which, with allowances for transport, would be around \$370/ton cif), the West German distributor could not; but that difference is largely due to the strength of the pound, as well as the fact that Middle East postings for LPG have not (at least, at the time of writing had not) escalated with crude oil prices.

Gas manufacture. The increased availability of natural gas in Europe has drastically cut the use of LPG for enrichment of town gas, for peak-shaving and, as propane/air mixtures, for the supply of remote or localised areas. The quantity used fell from 1.5 mn tons in 1967 to less than 0.5 mn in 1979. Except, however, in France, where its use continues to decline, a plateau seems to have been reached.

Although it seems improbable that LPG will be used in the near future in Europe for the manufacture of so-called substitute natural gas (SNG). it is not impossible, given a favourable conjunction of natural gas and LPG prices. Such a conjunction might result from a failure of current plans for the importation of natural gas and/ or excessive price expectations on the part of its suppliers. coupled with a surplus of LPG on a world scale beyond its uses as a premium fuel and petrochemical feedstock. The possibility is examined here in the context of current expectations about the role of natural gas in Western Europe's overall energy consumption and likely availability.

According to submissions made in 1979 to the IEA by its member countries, and with allowance for France (on the basis of its latest energy plan) and Finland. consumption of natural gas in Western Europe is forecast as rising from 176 mn tons of oil equivalent in 1978 to 242 mn in 1985 and 263 mn in 1990. Production is expected to rise from 152.5 mn tons in 1978 to 158.5 mn in 1985 and fall back to 143 mn in 1990. Details are shown in the table below. On this basis, the area's requirement for imported natural gas would rise from 23.4 mn tons of oil equivalent in 1978 to 83.5 mn in 1985 and 120 mn in 1990.

Forecast Production and Consumption of Natural Gas in Western Europe (mn tons of oil equivalent)

	1978	1985	1990
Production			
Austria	2.1	1.0	0.9
Denmark	-	2.0	3.0
France	6.9	5.0	2.8
West Germany	16.2	14.4	12.5
Greece	-	0.4	0.2
Ireland	-	1.1	1.1
Italy	11.3	10.0	10.0
Netherlands	68.6	62.6	52.7
Norway	14.2	23.0	18.0
Spain	-	1.7	1.8
UK	33.2	37.3[a]	40.0[a]
Total	152.5	158.5	143.0
Consumption			
IEA	152.4	205.6	220.0
France	22.6	35.5[b]	42.0[b]
Finland	0.9	0.9	1.0
Total	175.9	242.0	263.0
Import requirement	23.4	83.5	120.0

a Mid point of range. b Upper level of forecast.

Source: IEA, Comité Professionnel du Pétrole.

Although there are individual variations in the calorific value of indigenous and imported natural gas, ranging from a low of 8,400 kcal/m^3 in the case of Dutch gas, the above forecast is sufficiently widely framed for little distortion to result from taking the 1985 import requirement as equivalent to 83.5 bn cu metres of (10,000 kcal/m^3) gas and the 1990 requirement as 120 bn.

In 1979 imports of natural gas into Western Europe amounted to 29 bn cu m, of which 4.27 bn came from Algeria and 3.6 bn from Libya in the form of LNG, whilst approximately 21 bn cu m was piped from the USSR. Potential additional sources of supply include the following.

    i.   Algeria, by pipelines currently being laid under the Mediterranean and planned, or as LNG, up to 50 bn cu m by 1990.

    ii.   The USSR. The volume committed for 1980 is 22-25 bn cu m. Pipelines from the Yamal Peninsula in Western Siberia could permit the level to rise by 40 bn cu m by 1985, according to offers now on the table - and if political circumstances allow. Russian sources have also claimed that the supply could reach 70-80 bn cu m/year by 2000.

    iii.   Nigeria has concluded agreements to supply 8 bn cu m/year as LNG by 1990. If the USA maintains its current restrictions on LNG imports, the volume could reach 16 bn cu m.

    iv.   There is a possibility of a West German consortium shipping LNG from Qatar's North West Dome field. That would not take place until after 1990.

It is expected that the supply from Algeria will reach 32 bn cu m by 1985, and it is here assumed that current negotiations with the USSR will be concluded in time to permit pipelines to be laid for a supply by then of around 20 bn cu m from Yamal, on top of current exports. With continuing shipments of LNG from Libya, the import availability would amount to around 75.5 bn cu m, leaving a gap of about 8 bn cu m. If this gap were met by the use of LPG as substitute natural gas, it would require nearly 7 mn tons of the product.

As regards 1990, the estimates submitted to the IEA appear to be on the cautious side insofar as North Sea production is concerned. Additional gas-gathering systems are planned in both the UK and Norwegian sectors. On the UK side, the system to which the government has already given its approval suggests that the upper level of forecast availability (the official range is 35-45 mn tons of oil equivalent) could be reached by 1990. The Norwegian government has so far taken no decision on the method of exploitation of a number of known and newly discovered, but still to be delineated, gas fields. On the whole, however, it seems unlikely that production will fall by 5 mn tons of oil equivalent between 1985 and 1990, as postulated in the submission to the IEA. The implied addition from these sources would reduce the 1990 import requirement to approximately 110 bn cu m.

However, even without allowing for increased North Sea production, with approximately 50 bn cu m available from Algeria in 1990, 60 bn from the USSR, and contracted LNG supplies from Libya, Nigeria, and possibly Qatar, any gap between supply and demand would disappear.

188

It must be emphasised that the purpose of making the foregoing assessments is largely theoretical. Unless there is a considerable world surplus of LPG, implying a price in the long term entirely determined by its calorific value, after allowing for the additional cost of storing and distributing it in comparison with those fuels with which natural gas also competes, the product is unlikely to be used as substitute for natural gas in Europe. It is not so much that plant would not be built to reform 7 mn tons of LPG into natural gas because the requirement for it seemed likely to disappear within five years. But plant will not be built until the respective supply/demand and long term price relationships between natural gas, LPG and other fuels become clearer.

No allowance is made here for the possibility that natural gas may be used in Europe for SNG. That conclusion may, however, need to be re-examined if the findings of the study indicate the likelihood of a long term surplus of LPG on a world scale.

Petrochemical use. What role LPG may play in the future as a feedstock for Western Europe's petrochemical industry has become a favourite theme at recent gas conferences. The debate arises from the conjunction of (a) the expectation that naphtha, Europe's traditional feedstock, will be in tight supply in the future and (b) the probability that the opposite will apply to LPG because of increased availability from the North Sea, and from the Middle East and Algeria. The implication of these features of supply is that the price of LPG will be attractive vis-à-vis that of naphtha, after allowing for the greater cost of handling LPG. Clearly, however, this price relationship may not obtain if there is a concerted move to replace naphtha with LPG. Secondly, there are doubts about the longevity of LPG supply from the North Sea and the security of the Middle East as a source, coupled with indications that the producers in the latter areas have high price expectations in regard to gas generally, both natural gas and LPG.

An additional consideration is the relative yield of products from the cracking of various feedstocks, in particular the availability of propylene vis-à-vis ethylene, given that demand for the former has historically risen, and in the future is expected to rise, more strongly than the demand for ethylene[1]. The respective yields of pyrolysis gasoline as a feedstock for aromatics plants represent a second consideration. Obviously the market requirements for co-products vary from producer to producer, and oil companies with separate LPG streams available from their refineries probably have greater feedstock flexibility, depending on plant location, than purely chemical companies.

The table overleaf illustrates the point in relation to a production of 450,000 tons/year of ethylene.

---

1 The development by Houdry of its Catofin process for the catalytic dehydrogenation of propane could potentially overcome this problem. It is claimed that the process permits 77 per cent by weight of propylene to be recovered from a propane feedstock compared with a 33 per cent yield by thermal cracking.

Typical Ethylene Cracker Yields in Relation to Feedstock
('000 tons/year)

	Medium range naphtha	Atmospheric gas oil	95 % pure ethane	60 % ethane 40 % propane	70 % propane 30 % butane
Hydrogen	15.9	16.2	30.7	27.1	18.3
Methane	230.9	215.1	63.5	109.0	297.5
Ethylene	450.0	450.0	450.0	450.0	450.0
Ethane	–	–	–	–	–
Propylene	195.3	185.2	22.0	54.9	174.1
Propane	12.5		5.5	27.2	12.3
Total C4	120.0	99.0	11.2	16.9	53.1
Pyrolysis gasoline	243.3	297.7	18.2	28.2	69.0
Fuel oil	69.8	280.7	–	–	8.5
Feedstock	1,337.7	1,543.9	601.1	713.3	1,082.8

Past fears about the availability of naphtha have led to the incorporation in certain crackers, representing a little over one third of total nameplate capacity (currently 17.4 mn tons), of flexibility to process a proportion of gas oil; about 4 mn tons of gas oil is reported to have been used in 1978 out of a total throughput of 40 mn tons. There are few plants designed for naphtha/LPG flexibility or to process natural gas liquids. The major exceptions are, in the former category, the new ICI/BP cracker at Wilton (capacity 500,000 tons/year), which can handle a feedstock of up to 50 per cent LPG and, in the latter category, the Rafnes plant in Norway (capacity 300,000 tons/year), designed to use Ekofisk ethane/LPG returned from the pipeline to Teesside. Of plants currently proposed, those of Esso at Moss-moran and, later, Dow (or some other company) at Nigg Bay will use North Sea ethane as feedstock (the Esso plant will be able to use some propane); the LPG from the associated gas stream pipelined to shore will be largely exported. Dow has stated that it is intending to export LPG from Scotland to its proposed ethylene cracker at Tarragona in Spain (350,000 tons/year capacity, scheduled for completion in 1983/84) and to replace at least a part of the 3 mn tons/year of naphtha used at its Terneuzen, Holland, plant (capacity 900,000 tons/year); the company is also reported to be investing $50 mn to expand its LPG storage installations at the latter site. In Norway, both Norsk Hydro and Saga Petroleum are proposing to build ethylene plants (total capacity 0.8 mn tons/year) to use NGL from the Statfjord field, whatever decision is taken by the Norwegian government on the evacuation facilities for the field (i e if the pipeline landing is not in Norway, the NGL will be transported back, as in the case of NGLs from Ekofisk).

In addition, there are no doubt unreported projects for adaptation of existing crackers to use a proportion of LPG. It is technically possible to use up to 5 per cent butane as feedstock in most naphtha crackers whilst, according to Chemical Age, some producers have found that "only minimal modifications are required in order to run 10 or even 15 per cent". It is presumably on some such basis as this that one recent report envisages the use of LPG in the Rotterdam area alone rising from 1 mn tons in 1979 to 1.9 mn by 1985, (for both gasoline blending and petrochemical use) and to 3.5 mn by 1990. An LPG pipeline to DSM's complex at Rijnmond is projected in connection with this forecast, with plants being adapted to use up to 20-25 per cent LPG feedstock and, finally, a new cracker based entirely on LPG. The publication of this report, prepared by the Rotterdam port authorities, the Rijnmond area authorities and provincial councils, coincided with the granting of permission for a new LPG terminal in Amsterdam and with Shell/BP's application (since approved) to build a terminal in Rotterdam to handle an initial 2 mn tons/year of LPG (1984/85), and an eventual 4-5 mn tons. Finally, Shell has forecast[1] that LPG's share of feedstock for olefin production in Western Europe could rise from 6 to 9 per cent over the period 1978-90 - although it is not made clear in the forecast what quantity of LPG that might represent.

The variables that have to be taken into account in making a forecast are, in fact, formidable. Apart from the price relationships between the various feedstocks, there is the question of the future demand for ethylene and its co-products in Europe. Only on one point does there seem to be consensus, namely that current projects for additional plant imply that there will be no shortage of capacity to meet the demand, whatever it may be.

---

1 Oil and Gas Journal, November 26, 1979.

According to the Council of the European Federation of the Chemical Industry (Cefic), effective ethylene capacity in Western Europe amounted in 1979 to 14.7 mn tons - compared, that is, with a nameplate capacity of nearly 17 mn tons. Consumption in Western Europe - which, given the economics of long distance transport of ethylene, may be taken as commensurate with production - was 12.16 mn tons, equivalent to a utilisation of nearly 83 per cent of effective capacity. Cefic forecasts an increase in consumption averaging only 2.7 per cent in 1979-84, with the level reaching 13.9 mn tons by the end of the period. In terms of forecasts by other organisations and individuals, this represents a very low growth rate - pessimistic also in terms of Cefic's 1979 forecast, which postulated a 4.2 per cent increase in 1978-83. Historically the demand for ethylene has risen at a rate well in excess of GDP, though inevitably it will fall as the scope for substitution diminishes; it rose by over 10 per cent a year in 1969-79, associated with a GDP growth of 3.2 per cent, and by 6 per cent in 1973-79, when GDP growth averaged 2.6 per cent. It is here assumed that the growth rate will be in the range 3-4 per cent throughout the period 1979-90, on the assumption of a GDP rate of 2 per cent in 1979-85 and 2.5 per cent thereafter. That would imply a required ethylene production of 14.5-15.4 mn tons in 1985 and 16.8-18.7 mn tons in 1990.

On present evidence it seems unlikely that the various proposed ethane-based petrochemical plants in Scotland and Norway will be on stream by 1985, with the exception of Mossmoran. Required ethylene production from crackers based on naphtha/gas oil/LPG will then be of the order of 13.8-14.7 mn tons. By 1990, ethane-based capacity in Scotland and Norway could amount to 2.1-2.3 mn tons of ethylene. Allowing for a high utilisation of this new capacity, required production from other crackers would amount to 14.9-16.8 mn tons.

Excluding from both the feedstock and output sides of the equation the 550,000 tons of LPG that, according to Norwegian sources, was used in the Rafnes plant, LPG seems to have accounted for nearly 7 per cent of cracker feedstock in 1979 - around 2.6-2.7 mn tons. Taking a 7.5-10 per cent range for 1985 and 1990, and applying it to the mean of requirement as postulated above, would imply use of LPG in the former year of 3.2-4.2 mn tons (mean of 3.7 mn) and in the latter year of 3.5-4.7 mn tons (mean of 4.1 mn) - to which must be added back the consumption at Rafnes.

Consumption of LPG by the chemical, as opposed to the petrochemical, industry is not significant, amounting to 0.2-0.3 mn tons in 1979. There will be an increased demand for butane in the future for MTBE and TBA, raising the total to perhaps 0.5 mn tons in 1985 and 0.6 mn in 1990.

On this basis the requirement for LPG in the chemical/petrochemical industry would by 1985 be of the order of 4.8 mn tons and by 1990 around 5.3 mn tons. These respective means would imply growth rates in 1979-85 of 6 per cent, compared with 7.8 per cent in 1973-79, and in 1985-90 of 2 per cent. The lower rate in the latter period is essentially due to the increased use of ethane.

Summary

The table summarises the findings of the above sections in order to arrive at an estimate of potential imports into Western Europe of LPG from outside the area.

192

Western Europe: Potential Imports of LPG

('000 tons)

	1979	1985	1990
Indigenous supply	15.19	17.8-18.8	19.2-22.0
Demand:			
traditional uses	11.66	13.0	14.9
motor fuel	1.38	2.5	4.0
gas making	0.49	0.5	0.5
chemical/petrochemical	3.39	4.8	5.3
Total	16.92	20.8	24.7
Net imports	1.73[a]	2.0-3.0	2.7-5.5

a To meet observed consumption. According to OECD net
imports amounted to 1.99 mn tons (see table on page 3).

CHART 3.

USA: LPG Imports

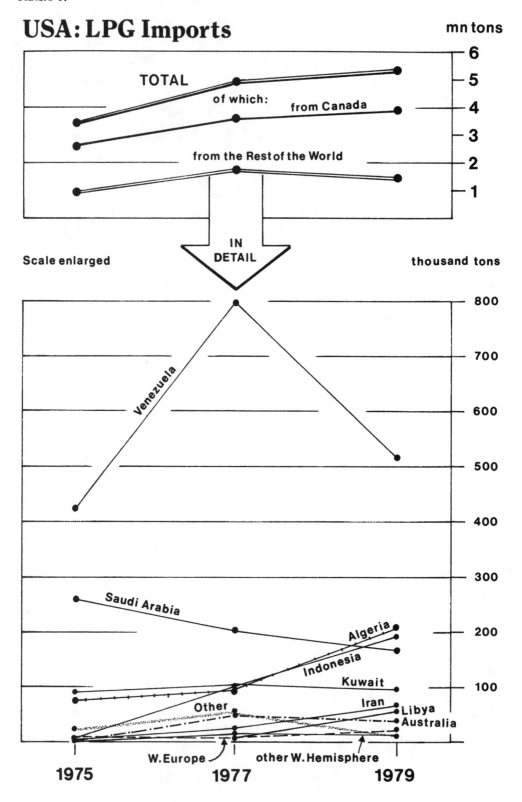

mn tons

Scale enlarged

thousand tons

CHAPTER 2

USA

Domestic supply of LPG

The first table below summarises the information on available supply of LPG in the USA for meeting domestic requirements in the period 1973-79. The second table details the source of imports in recent years.

US Production and Imports/Exports of LPG
(mn tons)

	1973	1975	1977	1979
Gas processing plants	28.42	26.95	25.00	24.74[a]
Refineries	10.53	9.03	10.04	10.20
Total production	38.95	35.98	35.04	34.94
Imports	4.06	3.45	4.96	5.35
Exports	0.85	0.80	0.55	0.45
Available supply	42.16	38.63	39.45	39.84

NB: US Bureau of Mines figures for propane and butane have been converted to metric tons at 12.35 barrels = 1 ton propane, 11.25 barrels = 1 ton butane, and 11.8 barrels = 1 ton propane/butane mixture.

a On assumption that ethane/propane mixtures contain 30 per cent propane in volume terms.

USA: Sources of LPG Imports
('000 tons)

	1975	1977	1979
Canada	2,567	3,609	3,937
Mexico	–	–	28
Venezuela	421	798	513
Other Western Hemisphere	2	31	–
Western Europe	13	19	34
Iran	7	25	67
Kuwait	80	108	91
Saudi Arabia	257	203	168
Other Middle East	3	34	3
Algeria	76	84	212
Libya	–	15	57

(continued)

USA: Sources of LPG Imports (continued)
('000 tons)

	1975	1977	1979
Other Africa	6	12	–
Indonesia	5	99	192
Australia	–	47	38
Other Far East	8	3	9
Other	6	–	–
	3,451	5,087	5,350

Sources: 1975 and 1977 OECD; 1979 US trade statistics.

Gas processing plants. In the past the extraction of LPG from natural gas has represented a remarkably constant 5.5 per cent of the thermal content of the gas processed. It is assumed that this will remain true in the future. The forecast of natural gas production submitted by the General Accounting Office to the US Congress in December 1979 envisages an increasing - though by 1990 still relatively small - contribution from the Outer Continental Shelf, where the gas is expected to be drier that that from currently produced, shallower formations. In estimating future production from gas processing plants (see table below), no allowance has been made for this factor, not least because the GAO forecast for output from the lower 48 states and Alaska seems anyhow to be on the pessimistic side. In total the forecast postulates a fall in production from 19.7 trillion cu ft in 1979 to 17.1 trillion in 1985 and 16.8 trillion in 1990. On the assumptions as set out above, the yield of LPG would be 21.55 mn tons in 1985 and 21.15 mn in 1990, compared with an estimated 24.74 mn tons in 1979.

Refineries. Output of LPG at US refineries has gradually declined as a proportion of crude throughput, from 1.7 per cent by weight in 1973 to 1.4 per cent in 1979, partly as a result of the heavier crudes being processed. For this reason, catalytic cracking capacity as a proportion of distillation capacity has been rising and will continue to rise, and will also be operated at high severities. The decline in the proportion of LPG is therefore likely to be stemmed, and might possibly be reversed. A range of 1.4-1.5 per cent has been applied in the table below to postulated crude throughput of 15 mn b/d in 1985 and 1990. The possibility of a fall in crude throughput to 13.65 mn b/d by 1990 has also been taken into account: if gasoline consumption in the USA falls as has been forecast by the National Petroleum Council, as discussed later, refinery throughput may be reduced to accommodate the fact.

Imports from Canada. Imports by pipeline from Canada averaged 127,840 b/d in 1979 (approximately 3.94 mn tons). They are forecast as peaking at 135,000 b/d in the early 1980s and declining to 95,000 b/d by the mid 1990s. This forecast does not take account of possible additional volumes available from Canadian Arctic gas or gas discovered offshore from Newfoundland - unlikely before 1990 and, in any case, probably low in LPGs.

The table below sets out the forecast of available supply in the USA from these three sources. It is also possible that some LPG may arrive in the USA by pipe-line from Mexico, but as the total quantity likely to be exported from Mexico in the future is itself uncertain, no attempt is made to distinguish between pipeline and seaborne supplies.

Forecast of LPG Supply from Indigenous Sources and Canada
(mn tons)

	1979	1985	1990
Gas processing plants	24.74	21.55	21.15
Refineries	10.20	10.50-11.25	9.55-11.25
Pipelined from Canada	3.94	4.10	3.50
Exports	0.45	0.20	0.20
Total	38.43	35.95-36.70	34.00-35.70

Demand for LPG

The table below shows in tonnage terms the consumption of LPG in particular uses in individual years during the period 1969-79. Any difference between consumption and available supply, as set out in an earlier table, represents stock changes and statistical errors either in the original data or resulting from the conversion from barrels to tons.

In 1979 US consumption of LPG regained the level it had reached in 1973. Within this overall picture, trends differed between the two broad categories of consumption, LPG as fuel and LPG as chemical feedstock: consumption increased modestly in the former and,until 1979, declined substantially in the latter. Unlike the present situation in Europe, where a potential shortage of naphtha seems to open the way for LPG and ethane being substituted for naphtha as a petrochemical feedstock, since the late 1960s US producers have been moving in the opposite direction, building plants to handle naphtha/gas oil feedstocks in the expectation of a shortage of LPG, as the domestic output of natural gas fell. The extent of the substitution that has taken place can in part be seen from the second table (which relates to the chemical industry as a whole).

US Consumption of LPG by End-Uses, 1969-79
(mn tons)

	1969	1973	1975	1977	1978	Preliminary 1979
Residential/ commercial	15.00	15.40	13.74	13.89	13.48	na
Automotive	2.87	2.77	2.28	2.22	1.92	na
Industrial	1.86	2.15	2.12	2.02	2.60	na
Utility gas	0.36	0.68	0.79	1.09	0.96	na
Gasoline production	6.47	7.13	7.97	7.57	7.76	7.61

(continued)

US Consumption of LPG by End-Uses. 1969-79 (continued)

(mn tons)

	1969	1973	1975	1977	1978	Preliminary 1979
Other (chiefly agriculture & SNG)	0.42	1.03	2.66	3.88	3.25	na
Total fuels	26.98	29.16	29.56	30.67	29.98	32.69
Petrochemical feedstock	10.23	11.70	8.17	7.20	7.28	8.50
Total	37.21	40.86	37.73	37.87	37.26	41.19
of which:						
propane	23.79	26.02	23.97	25.35	22.99	26.44
butane	10.93	12.87	12.41	11.55	13.18	13.70
propane/butane mix	2.49	1.97	1.35	0.97	1.09	1.05

NB: US Bureau of Mines figures for consumption in individual categories other than petrochemicals and gasoline production have been converted to tons on the basis of the relationship in each year of butane to propane in the total for those categories. using 12.35 barrels = 1 ton propane, 11.25 barrels = 1 ton butane, 11.8 barrels = 1 ton propane/butane mix. Gasoline production is taken as consisting entirely of butane, while petrochemical consumption is directly derived from the volume of each fraction used.

Chemical Feedstocks in the USA

('000 b/d)

	1973	1975	1977	1978	1979
Ethane	327	341	412	433	486
LPG	389	272	233	232	(270)
Oil fractions	356	320	521	595	674
Total	1,072	933	1.166	1,260	1,430
Oil as %	33.2	34.3	44.7	47.2	47.1

( ) Partly estimated.

Source: USBM.

It is estimated that whereas by end 1970 a quarter of US capacity for ethylene production was designed for naphtha/gas oil feedstock, the proportion rose to 35 per cent by end 1975 and to 50 per cent by end 1980. The last LPG cracker was brought on stream by Phillips in 1978. No more are currently planned. The question of particular relevance to this study is whether that trend will now be reversed. given a potential world surplus of LPG. The factor of price relationships between the two feedstocks is complicated by the expectation that naphtha may also be in potential surplus in the USA - if, that is, gasoline consumption falls. as has been widely forecast. without a commensurate fall in refinery runs.

LPG as fuel. Broadly speaking, since 1973 fuel use of LPG has increased significantly in only one sector. namely the production of utility gas and substitute natural gas (SNG). The total used for SNG is estimated to have reached 2.8 mn tons in 1977. accounting for nearly all the increase in the "other" category - whose remaining component is chiefly LPG for agricultural use.

As regards the future, it seems probable that the use of LPG for the production of SNG reached its peak in 1977. Of the 14 plants in existence (excluding one in Hawaii), the two latest were designed to process naphtha only, conforming to the Federal Energy Agency's dictum that to convert LPG to SNG was a wasteful use of a clean gaseous fuel. It is intended that future SNG production should be based on coal. In terms of installed capacity to produce SNG, there is an equal split between naphtha feedstock and NGL (chiefly propane, but also some ethane and natural gasoline), at 168,500 b/d each. Since the plants operate for only about half the year (November to March), the potential maximum use of LPG is around 2.5 mn tons/year. Further increase is, however, possible in the production of peak-shaving utility gas from LPG.

Whether consumption of LPG in the residential/commercial, industrial and agricultural sectors will rise in the future is in part dependent on the availability of natural gas and electricity in particular areas of the USA and in part on the interrelationship of fuel prices, specifically the price of propane vis-à-vis distillate fuel oil (No 2 fuel).

In the 1960s the price of propane at Gulf terminals (inlet to pipelines) was approximately equal in calorific terms to the average price of crude oil. From around 1969, the price rose relative to the price of crude oil, gradually approaching that of No 2 fuel oil with which it was more or less equated from 1973 onwards, sometimes rising 5-10 per cent above, as was justified by its clean properties in premium uses, sometimes undercutting. The table below indicates the recent change, as a result of controls being maintained on propane prices (controls were lifted on butane prices at the beginning of 1980, when, however, the margins of propane marketers were frozen at their May 1973 levels), while No 2 fuel oil has reflected the full rise in the average cost of US crude throughput in refineries.

Average Prices of Propane and No 2 Fuel, US Gulf
(cents/1,000 kcal)

	Propane at Mont Belvieu terminal	No 2 fuel Gulf spot	Contract cargo
Sep 1978	0.96	1.04	1.05
Mar 1979	0.93	1.47	1.25
Dec 1979	1.59	2.48	2.22
1980			
Mar	1.86	2.13	2.27
Jun	1.88	2.20	2.31
Sep	1.84	2.18	2.33

Source: Platt's Oilgram.

As was shown in an earlier table, there was in 1979 a substantial upturn in US consumption of LPG as a fuel. The table above suggests one of the reasons, namely the relative fall in price. The question is whether the recovery can be maintained. Those who have been arguing for decontrol of propane prices have pointed out that butane prices actually fell following decontrol, adding for good measure that there will be, in the future, no world shortage of propane: it is accepted that US production will decline, but production elsewhere, it is said, will be in surplus. Thus, for example, the National LP-Gas Association has forecast that demand for propane (including its use as a petrochemical feedstock, which, however, the association expects to decline) could rise by an average 1.2–1.3 per cent a year in the six years 1979–85, with imports doubling from 100,000 b/d to 199,000 b/d. The point, however, about imported propane – at least for seaborne supplies – is that its price is well above that of domestic and Canadian propane: the fob price posted by Kuwait and Saudi Arabia in May–September 1980 was the equivalent of 2.66–2.69 cents per 1,000 kcal. compared with 1.84–1.88 cents for propane at Mont Belvieu. The table below shows the average cif values of imported LPG in January–June 1980.

US Imports of LPG, January–June 1980

Source	'000 b/d	$/ton[a] Fas	Cif	Cif cents/ 1,000 kcal[b]
Canada	115.3	– 188.7	–	1.59
Mexico	13.3	235.5	266.9	2.25
Netherlands Antilles	0.6	255.7	281.7	2.38
Venezuela	12.6	318.7	350.1	2.96
UK	2.5	375.8	429.3	3.62
Saudi Arabia	4.2	278.7	333.7	2.82
Indonesia	2.1	401.0	421.8	3.56
Australia	0.5	265.0	413.7	3.49
Libya	0.8	319.1	375.5	3.17
Others	0.0	217.6	222.2	1.88
Total	151.9	213.2	222.2	1.88

a Barrels have been converted to tons at the 1979 import factor of 1 ton = 11.935 barrels (i e approximately 62 per cent propane and 38 per cent butane in the total). b Calorific content is taken as 11,845 kcal/kg on the basis of the above.

In brief, it was only because of the dominant weighting of Canadian supplies in total imports that the average price of the latter was, in calorific terms, below that of No 2 fuel in the first six months of 1980. The average value of seaborne imports (excluding Mexico) was in fact 3.06 cents/1,000 kcal.

In October 1980 the Saudi propane price was reduced to $305/ton (2.62 cents/1,000 kcal) and, it was reported, 30,000 b/d was being flared because of a lack of a market. In November, Kuwait was said to be requiring French would-be purchasers of additional supplies of crude oil to lift propane as well (80,000 tons of propane with 30,000 b/d of crude). An increase in US consumption based on imports has presumably been forecast in the expectation that such activities (flaring and composite sales: there is a limit to the amount that can be stored) will be forced on Middle East producers unless they reduce their price expectations about LPG. There is undoubtedly scope for the average US propane price to rise to the level of No 2 fuel, or even above it in certain specialist uses (glass making, textiles, food, certain iron and steel milling and finishing processes); and that could be achieved by rolling in import with domestic prices by propane producers. However, it is not at all clear that consumption would be greatly stimulated by this device. The evidence of 1973–78, when, as mentioned earlier,
200

the price of propane was more or less on a par with that of No 2 fuel, suggests that in the residential/commercial, automotive, industrial and agricultural sectors, taken as a whole, demand for LPG might be static. There would, at least, be little incentive for users in these sectors to convert to propane in a situation where (a) there is uncertainty about the price of domestic propane now that it has been decontrolled[1], and (b) in a context of a declining domestic supply, the weighting of higher cost imports from the Eastern Hemisphere would be increasing.

On this basis it is here assumed that, at best, LPG consumption in the fuel sectors (excluding gasoline production) might increase by 2 per cent a year in 1979-90 - compared, that is, with 2.2 per cent in 1973-79, but allowing for a decline in GDP growth. At worst, the level might fall back to the pre-1979 figure of 22-23 mn tons. A narrower and "most probable" range is assumed of 25-27 mn tons in 1985 and 27-29 mn tons in 1990.

Any increase in the use of butane as a gasoline blending agent depends on whether or not gasoline consumption indeed reached its peak in 1979, as seems to be widely believed in the USA. In the past it has regularly been forecast that vehicle owner-ship in the country would soon reach saturation point, and that gasoline consumption was approaching a plateau. Consumption did in fact dip after the 1973/74 oil price increases (by 4 per cent in 1974), but by 1976 the level was well above that of 1973, and thereafter it continued upwards until stemmed by the shortages and price rises of 1979. Between January-August 1979 and 1980 there was a more sizeable fall, of the order of 7 per cent. Continuing price rises, and the increased use of smaller cars, have prompted the National Petroleum Council to forecast a fall in consumption to 6.7 mn b/d by 1985 and 6.1 mn by 1990, compared with 7.03 mn in 1979. The Petroleum Industry Research Foundation foresees a more dramatic decline, to around 80 per cent of the 1979 level by 1990, implying no more than 5.6 mn b/d by then. The authors of this report are by no means certain that declines of that order will take place. In any case, in forecasting the requirement for butane for gasoline production, the probability of increased domestic self sufficiency has also to be taken into account: net imports of motor gasoline were of the order of 3 mn tons in 1979 (65,550 b/d).

In 1979 motor gasoline production represented 43 per cent of a crude throughput in US refineries amounting to 14.24 mn b/d (i e excluding NGLs and other hydro-carbon feedstocks from both sides of the equation). It was earlier postulated that crude throughput in US refineries might rise by a little over 5 per cent by 1985, to 15 mn b/d, and then flatten out. Output of gasoline would, at the 1979 crude relationship factor, amount to 6.45 mn b/d. If use of butane for blending rose pari passu, it would imply a requirement of 8 mn tons. The implication is that if the NPC forecasts of gasoline consumption in 1985 and 1990 are realised, the USA would be approximately self sufficient in the product in 1985 and would, by 1990, be a net exporter to the extent of 0.6 mn b/d. Alternatively, approximate self sufficiency in 1990, at a consumption level of 6.1 mn b/d, would imply a crude

---

1 Under President Carter's policy, the US crude oil price was scheduled to reach parity with world prices in late 1981 while deregulation of natural gas prices was to be postponed until 1985, applying then only to new gas. President Reagan has brought forward the former policy and also decontrolled propane prices.

throughput of 13.65 mn b/d, motor gasoline output of 5.87 mn b/d, and a butane requirement of 7.3 mn tons.

LPG as petrochemical feedstock. At the 1979 Gastech conference, a Union Carbide spokesman postulated, on the basis of a projected growth in ethylene demand, that the requirement for LPG as a feedstock in the US petrochemical industry could potentially reach 16 mn tons by 1985 and 18.4 mn by 1990. His qualifications to the forecast were as follows:

"Looking closely at the US petrochemical demand for LPG, we find that the basic demand - that is, the requirements of LPG based plants - will remain quite level through the next decade (at around 8.5 mn tons). There will be essentially no net change in the need for LPG as a primary feedstock. However, we anticipate a potentially substantial increase in demand to meet the appetite of naphtha/gas oil crackers that have some LPG feedstock flexibility. The (postulated) overall increase in LPG demand is therefore entirely due to this new trend of building into olefins units a margin of flexibility that permits some LPG to be processed in crackers designed primarily for other feedstocks.

"Expanding the margin of feedstock flexibility in petrochemical facilities has become an important planning strategy for this industry over recent months. We now estimate the industry can convert between 15 and 20 per cent of its naphtha feedstock base to LPG.

"We feel, however, that there is a 'natural' limit to the extent to which LPG can be used in naphtha or gas oil crackers. A major limiting factor would be the demand for co-products from naphtha or gas oil, such as aromatics, for which consuming and processing units have already been built in tandem with the cracking unit. LPG can be substituted only when production of those co-products may be sacrificed. Another constraint would be the special facilities required to handle and store large quantities of LPG, were it to become the primary feedstock. And an important consideration is always the relative price of one feedstock over another, in part based on comparative by-product values."

(To illustrate this final point: for a plant in the USA with flexibility to substitute a certain proportion of LPG for gas oil, the break-even point at third quarter 1980 US prices for co-products was approximately represented by parity between LPG and No 2 fuel oil prices on a weight basis. The fact that, at that time, the Mont Belvieu price for propane was around $220/ton compared with a spot price for No 2 fuel oil of $236/ton and a Houston contract price of over $250, in part explains the reported increase in use of LPG last year. In fact, as noted earlier, LPG use had already begun to rise in 1979.)

Demand for ethylene in the USA rose during 1959-69 at an average annual rate of 13 per cent. In the decade 1969-79 the growth rate fell by half to 6.5 per cent. Various forecasts have been made for the 1980s, of which perhaps the most modest is Shell's: a rate 2 per cent higher than GNP growth - or 3.6-4.8 per cent a year during 1979-90. A BF Goodrich forecast is somewhat higher, and more in line with Union Carbide's: around 6 per cent a year in 1979-84, falling to 4 per cent in 1984-89.

202

CHART 4.

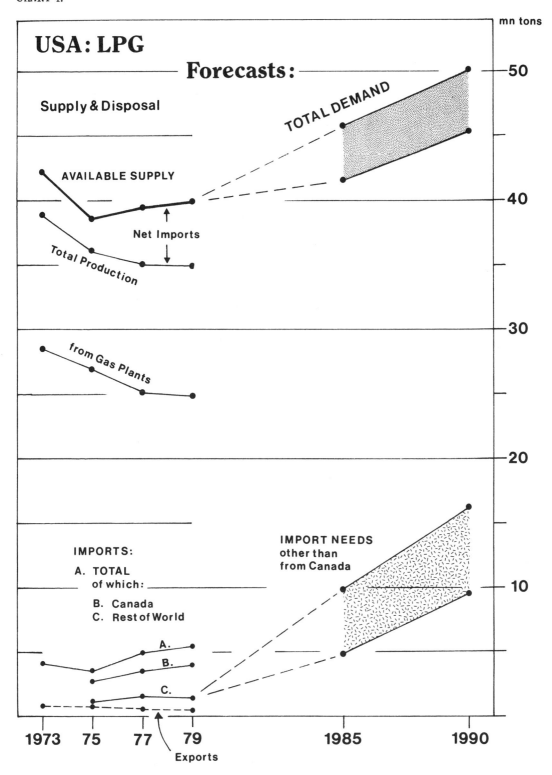

USA: LPG

Forecasts:

Supply & Disposal

TOTAL DEMAND

AVAILABLE SUPPLY

Net Imports

Total Production

from Gas Plants

IMPORTS:
A. TOTAL
   of which:
   B. Canada
   C. Rest of World

IMPORT NEEDS
other than
from Canada

A.

B.

C.

1973 75 77 79   1985   1990

Exports

mn tons
—50
—40
—30
—20
—10

For the purposes of this study it is assumed that GNP in the USA will rise by only 2 per cent in 1979-85 and by 2.5 per cent in 1985-90, and that ethylene demand will increase by 4 per cent a year in both periods. In so far as LPG is concerned (including a relatively insignificant use for other chemical processes – chiefly synthetic rubber), a range is postulated: a base demand of 8.5 mn tons in 1979-85, which is equivalent to an assumption that the use of LPG in 1979 represents a maximum in terms of plant capacity and flexibility in that period but with increased flexibility being built into plant coming under operation after 1985; and an increase in LPG use rising pari passu with ethylene production. On these assumptions demand for LPG as a petrochemical feedstock would rise as follows:

	1979	1985	1990
Mn tons	8.5	8.5-10.8	10.8-13.1

Summary

The forecasts in the previous sections on US supply of and demand for LPG are summarised in the table below to arrive at potential imports in 1985 and 1990.

Potential US Imports of LPG
(mn tons)

	1979	1985	1990
Indigenous supply[a]	39.1	36.0-36.7	34.0-35.7
Demand:			
gasoline production	7.6	8.0	7.3-8.0
fuel uses	25.0	25.0-27.0	27.0-29.0
feedstock	8.5	8.5-10.8	10.8-13.1
Total	41.1	41.5-45.8	45.2-50.1
Import needs	1.4[b]	4.8-9.8	9.5-16.1

a Including imports from Canada and, in 1979, stock reduction of 0.7 mn tons. b Actual seaborne imports.

204

CHAPTER 3

Japan

The historical situation

According to figures issued by the Ministry of International Trade and Industry, inland sales of LPG in Japan rose from 11.63 mn tons in 1973 to 15.25 mn in 1979, representing an average annual growth rate of 4.6 per cent. Imports grew at a rate of nearly 11.2 per cent a year, from 5.03 mn tons in 1973 to 9.49 mn in 1979. Throughout the period, Japan has represented the world's largest single market for imported LPG, concluding contracts with new producers well before their plants came on stream, partly as an aspect of a policy to diversify energy sources, partly as a result of preoccupations with atmospheric pollution. The process continues, with MITI forecasting that imports of LPG will once more nearly double in a period of six years, 1979-85.

Japan: Sources of LPG Imports
('000 tons)

	1975	1977	1979
Australia	1,129	1,177	1,395
Indonesia	-	-	84
Other Far East	45	62	15
Kuwait	771	1,036	1,841
Qatar	104	57	-
Saudi Arabia	2,490	3,884	5,297
UAE	-	31	489
Iran	678	751	126
Other Middle East	148[a]	-	-
Algeria	53	-	-
Canada	195	222	245
Venezuela	30	48	-
Other	37	12	-
	5,680	7,279	9,491

a Neutral Zone, i e Kuwait and/or Saudi Arabia.

Sources: OECD and MITI.

MITI's forecast

The Ministry of International Trade and Industry last year revised its long term forecast for the country's energy consumption, postulating a growth rate of 4.8 per cent a year in the period 1978-85, falling to 4.4 per cent in 1985-90. These energy rates are associated with forecast GDP growth rates of 5.5 and 5 per cent respectively, contingent upon the energy coefficient being reduced to 0.89 compared with unity in 1960-73 and 0.32 in 1973-78.

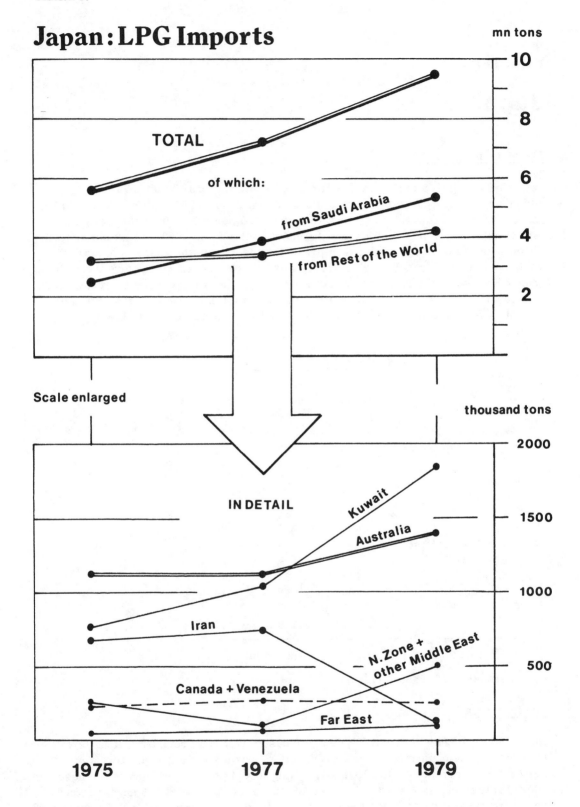

CHART 5.

Japan: LPG Imports

mn tons

TOTAL

of which:

from Saudi Arabia

from Rest of the World

Scale enlarged

thousand tons

IN DETAIL

Kuwait

Australia

Iran

N. Zone + other Middle East

Canada + Venezuela

Far East

1975   1977   1979

The forecast envisages a major replacement of oil by nuclear power, imported coal and LNG, and new energy forms, permitting net oil imports to be held at 6.3 mn b/d (325.7 mn tons), a level expected to be reached in 1985. Thus, oil imports would grow during 1978-85 from a base level of 5.4 mn b/d (266 mn tons) at an average annual rate of 2.9 per cent, and then flatten out. In connection with the proposed substitution, the IEA has observed: "If these goals are not achieved – and there must be grave doubts about those for coal, LNG and nuclear power generation – Japan will have no other alternative than a lower economic growth rate, a major restructuring of industry, or higher oil imports than now planned".

The restructuring of industry is a continuous process in Japan, marked latterly by the acceptance of a shift of energy and labour intensive industries to lower cost countries in the developing Far East. If the effect of that on energy consumption is ignored, of the other two alternatives the first seems, at the present moment, the more likely. Between 1973 and 1979, both boom years for the Japanese economy, GDP rose at an annual average rate of 4 per cent. In the interim the world economy was recovering from the impact of the first steep Opec price rise. It has still to start its recovery from the second, and in at least one respect the situation is less favourable: in 1974-78 there was a continuous fall in the real price of oil; currently there is the prospect of a steady rise, stemming from Opec's announced intention to control oil availability. In so far as concerns LPG as an element in Japan's forecast of total energy imports, the factor of price is obviously relevant, though availability may not prove to be a constraint. In other words, LPG imports could increase as envisaged by MITI even though total oil imports may not.

MITI's forecast of LPG imports in fact extends only as far as fiscal 1984, as does its forecast of demand, except in respect of use of the product for electricity generation. The forecast to 1984 is as follows.

Japan: LPG Supply and Demand, Fiscal Years 1978-84
('000 tons)

Supply	1978	1980	1984
Ex refineries	4,279	4,234	4,854
Ex petrochemical plants	402	360	360
Ex oil/gas fields	21	20	20
Imports	8,232	10,400	17,252
Total	12,934	15,014	22,486
Demand			
Household/commercial	5,357	5,794	6,716
Industry	3,362	3,426	5,350
Power generation	344	700	2,969
Town gas	924	1,371	2,019
Motor fuel	1,725	1,789	1,961
Chemical feed	1,257	1,773	3,173

(continued)

Japan: LPG Supply and Demand, Fiscal Years 1978-84 (continued)
('000 tons)

Demand	1978	1980	1984
Exports	36	10	10
Stock increase	(71)	151	288
Total	12,934	15,014	22,486

NB: A certain confusion is introduced by the fact that MITI issues two sets of LPG statistics, the other of which includes total production of LPG from naphtha crackers on the supply side. That has to be compensated by the inclusion, on the demand side, of the use by petrochemical plants of their own LPG for fuel and feed. On that basis demand/supply in fiscal 1978 was equated at 14,524,560 tons. The above table includes only the LPG marketed ex petrochemical plants.

The projected increase in imports of LPG averages 13.1 per cent a year in 1978-84, compared with 11.2 per cent in 1973-78, while demand (excluding exports and stock changes - see footnote to table) is forecast as rising at 9.4 per cent a year compared with 5.6 per cent in the earlier period. The fact that the latter was associated with a 3.6 per cent increase in GDP, whereas the forecast GDP growth rate for 1978-85 is 5.5 per cent, is not the justification for the higher growth rate. That is essentially posited on the increased availability of LPG relative to other oil products, and its substitution for them in large-consumer end-uses: specifically, for crude oil, fuel oil and naphtha in power generation, and for naphtha in the chemical and petrochemical industries and in the manufacture of town gas. In other sectors (those in which there is a multiplicity of small consumers), the projected growth rates are below or only slightly above those that obtained in 1973-78.

The purpose of looking at the sectors of expected higher growth rate is not to measure the economic realism of the forecasts. The realism depends on the availability of LPG to the world as a whole. To secure its projected share in that supply - given the price relationships between the various oil fuels - the Japanese government is already using and will increasingly use a range of tariff, duty and price measures, together with low interest loans channelled through various agencies. The seriousness of the endeavour is indicated by the fact that the total energy related budget - MITI's and those of the two special funds, the Coal and Petroleum/Non-Petroleum Alternative Energy Account and the Electric Power Resources Development Account - was increased by 31 per cent between fiscal 1979 and 1980, to Y729 bn ($3.3 bn at the June 1980 rate of exchange), while funds for the fiscal loan and investment programme handled by the Japan National Oil Corporation and the Japan Development Bank were raised by 26.5 per cent to Y382 bn ($1.7 bn).

Power generation

As regards power generation, although there is no legal prohibition, the government's basic position is that no more oil fired stations will be authorised - except for those already under construction or at the planning stage - whilst some will be switched from oil firing to gas. In that context the Utility Industry Council has forecast that consumption of LPG will rise from 344,000 tons in 1978 to 3.9 mn in 1985 and 4.1 mn in 1990, and of LNG from 8.6 mn tons to 21 mn in 1985 and

30-33 mn in 1990. The IEA considers it regrettable that a premium fuel like LNG should be used to the extent planned (which applies equally to LPG). but notes that it is justified by the Japanese authorities on the grounds of the short construction time for LNG power stations and the environmental benefits. (The latter is, in fact, a quite dominant consideration as a counterweight to the higher coal burn proposed - and one of the reasons why this part of the forecast is likely to be realised, with a special tariff being introduced as an incentive, as has already been done in industrial heat production.) The IEA also doubts that sufficient LNG will be available; but that only increases the attractiveness of LPG, should a world surplus of the latter emerge - as the Japanese authorities clearly expect.

The doubts of private industry in Japan about this programme seem to relate to the physical and time constraints (shortage of land for reception terminals and storage, residents' objections) and the long term security of LPG supply once power stations have been built for the purpose, rather than to the financial constraints - of which comparative fuel costs would. in other countries, be not the least important consideration. The table below indicates their significance in mid 1980.

Comparative Fuel Prices in Japan, June 1980

	Cif value of imports		Primary distributors' prices	
	$	US cents per 1.000 kcal	Yen	US cents per 1,000 kcal
LPG	351.67/ton	2.98	98,330/ton	3.79
Naphtha	39.11/barrel	3.00	60,000/kl	3.32
Heavy fuel oil C	29.07/barrel	1.87	63,200/kl	2.61-2.94
Crude oil	33.46/barrel	2.29	52,890/kl	2.62

NB: 1. The relationship between the cif values of LPG and naphtha, favourable to the former on a calorific basis, moves the other way when the higher costs of terminalling and distributing LPG is taken into account. 2. The cif value of heavy fuel oil C relates to a product of the average sulphur content of imports. It is comparable with the low of the distributors' price range: the high of the range relates to 0.3 per cent sulphur. high pourpoint fuel oil for thermal power stations. However, in the above calculations an exchange rate of Yen220 = $1 has been used throughout. Insofar as fuel oil for thermal power stations is concerned, the exchange rate officially used for April-June deliveries was Y244.56 (the average of first quarter import rates). which reduces the price to 2.64 cents per 1,000 kcal.

Town gas manufacture

In the case of the proposed increased use of LPG in town gas manufacture, the fuel to be replaced is naphtha, consumption of which is projected as falling from 2.08 mn tons in 1978 to 1.2 mn in 1984. As the table shows, the price relationship is here not so adverse. Whether the substitution is economically justified will in part depend on whether the Japanese government is correct in thinking that naphtha prices will rise, because of a long term shortage. relative to the price of LPG, which is expected to be in surplus. Clearly the relationship between the prices will also depend on the degree of substitution in Japan and elsewhere. In the first six months of 1980, the cif values of the two moved as follows.

Cif Values of Naphtha and LPG in Japan. January-June 1980

	Naphtha		LPG	
	$/barrel	Cents per 1,000 kcal	$/ton	Cents per 1,000 kcal
Jan	39.78	3.05	282.76	2.40
Feb	42.65	3.27	319.99	2.71
Mar	41.36	3.17	320.57	2.72
Apr	40.51	3.11	321.52	2.72
May	39.94	3.06	343.46	2.91
Jun	39.11	3.00	351.67	2.98

Allowing for terminalling and distribution costs, it would have made economic sense to substitute LPG for naphtha in town gas manufacture in January-February 1980, but not thereafter.

Petrochemical feedstock

The same price interdependence applies in the projected substitution of LPG for naphtha as a petrochemical feedstock. Thus, in early 1979 the Japanese petro-chemical industry was substituting LPG for naphtha, but later in the year the price of naphtha fell relatively to that of LPG, itself partly as a result of the lower level of capacity utilisation of naphtha crackers. The situation then switched back, but by mid 1980 had again changed - with, as shown in the table above, an increase between January and June of 24 per cent in the average value of imported LPG, coupled with a slight fall in the naphtha price.

The economics of substituting LPG for naphtha, and the extent of the ability to do so, vary from plant to plant, but on a broad rule of thumb basis, taking account of the different proportions of products yielded by LPG and naphtha (more ethylene and propylene from an LPG cracker, less pyrolysis gasoline and fuel oil), the break-even point is approximately represented by $35/barrel for naphtha and around $300/ton for LPG. In June 1980, the cif value of naphtha, at $39.11/barrel, represented a break-even price for LPG of $355/ton, compared with an average cif value for imported LPG of $351.67/ton. In terms, however, of delivered costs (i e taken from primary distributors), naphtha at Y60.000/kl ($43.36/barrel) represented a break-even LPG price of $371/ton, compared with a delivered price of $447. A persistence of such a relationship in the future would require the Japanese authorities to provide incentives for their proposed switch from naphtha to LPG.

Forecast supply/demand balance

In the context of the present study the question of whether the price of LPG is likely or not to move in a manner that would obviate the need for such fiscal/financial measures must be postponed until the total world supply/demand balance has been considered. In the meantime it is proposed substantially to use the Japanese official forecast in drawing up the demand side of the equation. As noted, this forecast does not envisage any increase in oil imports after 1985 (and hence, it must be assumed, in refinery throughput and availability of LPG from that source). By then also, the major switch to LPG will have taken place, at least insofar as concerns large scale end-uses; some continuing increase may be expected in the case of other uses, as a motor fuel, in industry and in household/commercial premises, depending on the growth of the economy as a whole.

210

Japan: Demand and Apparent LPG Import Requirements, Fiscal 1985 and 1990
(mn tons)

	1985	1990
Household/commercial	6.63- 6.97	7.85- 8.25
Industry	5.24- 5.78	7.45- 8.20
Power generation	3.20- 3.90	4.10
Town gas	1.96- 2.30	2.50
Motor fuel	1.94- 2.00	2.15- 2.20
Chemical feedstock	3.13- 3.80	4.00
Total demand	22.10-24.75	28.05-29.25
Domestic supply	5.35	5.35
Apparent imports	16.75-19.40	22.70-23.90

In the table above, the low of the range in fiscal 1985 and 1990 is based on an
assumption of an annual average increase in GDP growth rate of 4.5 per cent in
1978-85, and the high of the range on the assumption of the official forecast rate
of 5.5 per cent being attained. In respect of 1985-90, the official 5 per cent growth
rate is used.

Expressed in round figures, and converting the import totals in the table to
calendar years, imports would rise from an actual level of 8 mn tons in calendar
1978 (a level that went up by 18.6 per cent in 1979 to 9.5 mn) by 10.7-13 per cent
a year in 1978-85, to 16-19 mn tons at the end of the period, and by 4.3-6.3 per
cent a year in 1985-90, to 22-23 mn tons in 1990.

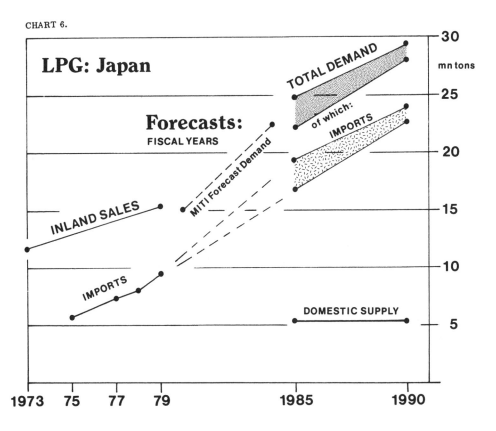

CHART 6.

CHAPTER 4

Rest of the World

Introduction

Data on countries other than OECD members are sparse. Such information as is available, deriving from the United Nations publication "World Energy Supplies 1973-78", is summarised in the table below. The chief interest inevitably attaches to the supply potential of Opec members as a group, and this question is discussed in the first section below. The communist bloc is broadly in balance, and for the purposes of this study it is assumed that it will continue to be so. Oceania is a net exporter by virtue of production in Australia, while the third world excluding Opec is a modest net importer. The latter areas are subsequently discussed in turn.

Rest of World: Production, Net Trade and Apparent Consumption of LPG ('000 tons)

	1973	1975	1977	1978
Africa				
Opec:				
production	600	864	844	1,010
net imports (exports)	(254)	(343)	(178)	(168)
consumption	346	521	666	842
Non-Opec:				
production	296	331	301	315
net imports	174	262	407	411
consumption	470	593	708	726
Central/South America				
Opec:				
production	2,320	1,851	2,131	2,206
net imports (exports)	(1,605)	(1,246)	(1,315)	(1,360)
consumption	715	605	816	846
Non-Opec:				
production	5,196	5,933	6,827	7,226
net imports	1,491	1,443	957	1,025
consumption	6,687	7,376	7,784	8,251
Middle East				
Opec:				
production	4,904	5,739	7,312	8,015
net imports (exports)	(4,319)	(5,094)	(6,702)	(7,385)
consumption	585	645	610	630
Non-Opec:				
production	650	650	704	736
net imports	28	160	248	248
consumption	678	810	952	984

(continued)

Rest of World: Production, Net Trade
and Apparent Consumption of LPG   (continued)

('000 tons)

	1973	1975	1977	1978
Far East				
Opec:				
production	8	28	45	45
net imports (exports)	(2)	5	(3)	(1)
consumption	6	33	42	44
Non-Opec:				
production	1,099	1,258	1,439	1,618
net imports (exports)	(53)	(38)	(56)	(39)
consumption	1,046	1,220	1,383	1,579
Oceania				
Production	1,343	1,615	1,799	1,833
Net imports (exports)	(1,038)	(1,073)	(1,363)	(1,370)
Consumption	305	542	436	463
Communist bloc				
Production	6,907	8,143	8,880	9,262
Net imports (exports)	(157)	(262)	(402)	(261)
Consumption	6.750	7,881	8,478	9,001

Source: United Nations.

The Opec exporters

Algeria. The following forecast was made of Algeria's production and utilisation
of LPG in a paper presented by a Sonatrach official in March 1979:

(mn tons)	1978	1979	1980	1985	1990
Production					
Oil fields/gas plants	0.66	2.60	4.50	8.40	7.90
Refineries	0.58	1.10	1.30	4.20	4.20
Total	1.24	3.70	5.80	12.60	12.10
Imports	0.18	–	–	–	–
Domestic use	0.63	0.73	0.88	0.99	1.70
Exports	0.79	2.97	4.92	11.61	10.40

In fact, in 1979 Algeria exported only 317,736 tons of LPG, 2.8 per cent above
the 1978 level, and this failure to realise targets has been so recurring a theme
in the country's oil history that there must be some doubt about the projected
1985 and 1990 export levels. The new government is, moreover, reviewing its
oil and gas options, the most publicised aspect of which has been the cancellation
of proposed LNG plants in favour of delivering gas to Europe by pipelines under
the Mediterranean. That need not affect plans for the production of LPG and
condensate in oil fields and gas plants, particularly since Algeria's limited oil
reserves mean that maintenance of oil exports are contingent upon increasing the
proportion of these products. In addition, LPG plants are for the most part
already under construction. On the other hand, considerable doubt must attach
to an expansion of the capacity of local refineries to the point where they would
yield 4.2 mn tons/year of LPG. The assumption is made here that, optimistically,
8.5 mn tons of LPG may be available for export by 1985, falling to 7.5 mn by 1990.

CHART 7.

Rest of the World: LPG Consumption

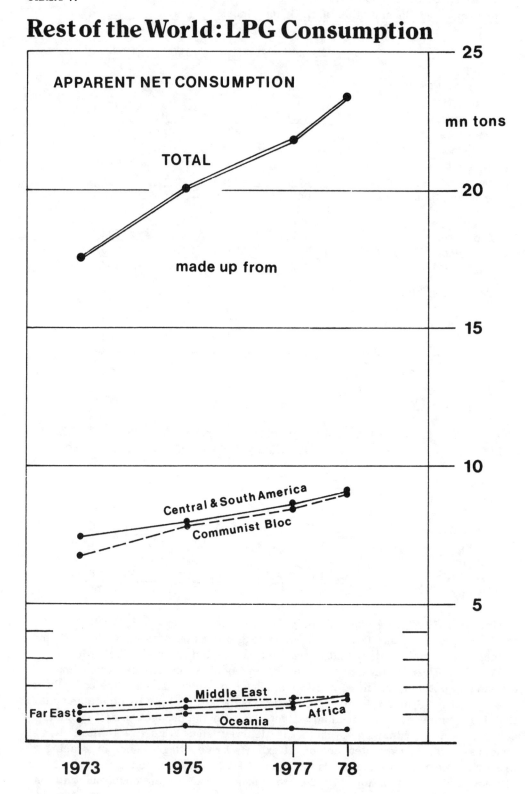

Ecuador. No plans for the export of LPG.

Gabon. Ditto.

Indonesia. The country's LNG export plants, based on non-associated gas at Arun and Badak, do not separate the LPGs, which are shipped in the LNG stream. (This is a growing practice, applying also in the case of existing and proposed LNG plants in Libya, Brunei and Malaysia; Australia, on the other hand, has decided against it.) Exports of LPG last year, derived from refineries and the NGL plant at the offshore Ardjuna field, are estimated to have amounted to 0.3 mn tons. In theory the volume could increase with utilisation of gas currently flared and available from new discoveries. But such a development is probably low down on Pertamina's scale of financial priorities. In fact, exports could dry up as a result of rising domestic demand, unless the proposed revamping/expansion of the Cilacap and Balikpapan refineries takes place and/or the Bataam refinery is built.

Iran. It was expected that by 1985 Iran would be exporting 2.5 mn tons/year of LPG from the Bandar Mahshahr NGL plant, coastal refineries, and the Bandar Shahpur petrochemical complex. That, however, was before the revolution and the war with Iraq. A quite arbitrary range of 0.5-1 mn tons is assumed for 1985, and 1-2 mn for 1990.

Iraq. The extent of the damage to Iraq's installations as a result of the war is also unknown. Even before that the various official announcements about LPG production and export were confusing. On the assumption that most of the refinery output of LPG will be consumed locally, and that most of the LPG from the NGL projects in the northern and southern oilfields will be exported, the quantity available is forecast at 4 mn tons in 1985, rising to 5-6 mn by 1990.

Kuwait. Production of LPG amounted in 1979 to nearly 2.8 mn tons, of which 2.75 mn was exported. The new Shuaiba plant has a theoretical capacity of 5 mn tons/year of NGL, including over 4 mn of LPG. Shortage of associated gas due to limitations on crude oil production (now 1.5 mn b/d compared with 2.23 mn in 1979) implies that it is unlikely to operate at even two thirds of capacity, unless non-associated gas is discovered in the current programme of deep drilling to the Khuff zone. Exports of 2-2.5 mn tons are assumed for 1985, possibly rising to 3 mn by 1990.

Libya has a theoretical capacity to produce about 1 mn tons/year of LPG from an NGL plant on the Intisar field and the Zawia refinery. Production last year is estimated at around 0.6 mn tons, of which about half was exported. The new Ras Lanuf export refinery will provide an additional 0.2 mn tons/year. Exports are forecast at 0.6 mn tons.

Nigeria. Until recently Nigeria has had to import LPG, but new refineries at Warri and Kaduna should meet domestic demand, and leave about 0.2 mn tons available for export in 1985. The proposed LNG export plant at Bonny (1.6 bn cu ft/day) would in theory make available around 1.3-1.6 mn tons/year of LPG, but the project has not advanced sufficiently for it to be known whether or not the LPG will be shipped in the LNG stream. LPG exports in 1990 are therefore postulated at 0-1.6 mn tons.

Qatar. The shaikhdom's first NGL plant, based on associated gas from the onshore Dukhan field, was destroyed by fire in April 1977. It is being rebuilt with a capacity of 1 mn tons/year. A second plant, utilising offshore gas, is expected to provide 0.7 mn tons of LPG for export by 1983. The plant has a capacity to produce 1,100 tons/day of propane and 900 tons/day each of butane and natural gasoline from 360 mn cu ft/day of feedstock. Available feedstock from the offshore fields of 250 mn – which is expected to decline to 190 mn by 1985 – implies that both plants will not be able to operate at full capacity unless additional supplies of suitable gas are located. A proposed export refinery of 150,000 b/d will yield approximately 0.1 mn tons of LPG. No allowance is made for LPG that might become available from an LNG project based on non-associated North-West Dome gas.

Saudi Arabia. According to official Saudi sources, LPG production in 1979 amounted to 79.5 mn barrels and domestic consumption to 2.7 mn, which suggests that LPG exports were nearly 6.6 mn tons. By 1983 installations will be in place at Ras Tanura, Juaymah and Yanbu for fractionating the NGLs stripped out of associated gas at the Berri, Shedgum and Uthmaniyah separation plants and contained in refinery streams available at the three ports. Exports are officially forecast at 14.5 mn tons of LPG and 5 mn of natural gasoline. Whether this level is in fact reached will in part depend on the level of crude oil production (and the ratios of light to heavy crudes in the total). Similarly, availability would be greater by 1990 if crude oil production were allowed to rise above 9.5 mn b/d. In any case, expansion of refinery capacity will lead to a greater availability of LPG: refinery capacity to meet local demand is expected to be 640,000 b/d by 1985 and for export 1.3 mn b/d; the former is forecast at 1 mn b/d by 1990 and the latter tentatively put at 1.8 mn. Thus, LPG export availability in 1990 could exceed the 14.5 mn tons postulated for 1985, despite an expected steep rise in domestic consumption.

UAE. The Dugas plant in Dubai is now in operation, exporting 0.6 mn tons/year to Japan. The Abu Dhabi Gas Liquefaction Company's plant on Das Island has a capacity to produce 0.8-1 mn tons/year of LPG along with 400 mn cu ft/day of LNG. It has not so far operated at more than half capacity. The Gasco NGL plant at Ruwais, based on onshore gas, will yield around 2.6 mn tons/year, and an export refinery at the same site a further 0.1 mn tons. Exports by 1985 may reach 3.5 mn tons.

Venezuela. Production of LPG from gas plants and refineries is of the order of 2 mn tons/year, with exports in 1979 of 32,800 b/d (1.03 mn tons). A separation plant under construction at San Joaquín in Anzoátegui and a fractionator at Puerto la Cruz, scheduled for completion in 1982/83, will increase capacity by 0.7 mn tons/year. Refinery yield will also increase with the coming on stream of catalytic crackers under the current up-grading programmes. Increasing domestic consumption will limit exports to around 1.5 mn tons in 1985. Utilisation of recent offshore gas discoveries may permit that export level to be maintained in 1990. Otherwise it will fall to 1 mn tons. Official forecasts are for exports of 47,000 b/d in 1985 (1.47 mn tons) and 34,000 b/d in 1990 (1.07 mn).

Opec. Summarising the above, the following table gives an estimate of the export availability of LPG from Opec member countries.

216

Opec: Export Availability of LPG

(mn tons)

	1979[a]	1985	1990
Algeria	0.3	8.5	7.5
Indonesia	0.3	0- 0.3	0.3- 0.6
Iran	0.3	0.5- 1.0	1.0- 2.0
Iraq	0.0	4.0	5.0- 6.0
Kuwait	2.8	2.0- 2.5	2.0- 3.0
Libya	0.4	0.6	0.6
Nigeria	-	0.2	0- 1.6
Qatar	-	0.8	1.2- 1.5
Saudi Arabia	6.6	14.5	14.5
UAE	0.4	3.5	4.1
Venezuela	1.0	1.5	1.1
Total	12.1	36.1-37.4	37.3-42.5

a Partly estimated.

Non-Opec importers/exporters

Central/South America.   Mexico has only recently become an exporter of LPG,
with a 1980 target in the range 30,000-40,000 b/d - say, 1 mn tons - from the
Pajaritos terminal.   An export terminal is also under development on the Pacific
coast, at Salina Cruz.   The amount of LPG available for export in the future will
be a function of the level of crude oil production that is permitted (a potential of
5-7 mn b/d by the end of the 1980s has been postulated, compared with 2.75 mn
by end 1980) and of domestic consumption, including Pemex's ambitious plans for
development of the petrochemical industry.   Conceivably it could be as much as
5 mn tons/year by 1990.

Of the importers, Brazil and Argentina are the most significant now that Mexico
has become an exporter.   Brazilian imports amounted to 76,000 tons in 1979, of
which 48,000 came from Mexico, 22,600 from Venezuela and most of the remainder
from Bolivia.   With utilisation of associated gas from offshore fields it is probable
that the country will become self sufficient.   The same is true of Argentina, where
proved reserves of gas have increased from 246 bn cu m at end 1977 to 600 bn at
end 1979; development of these gas resources, assisted by a World Bank team and
finance, is under way.   Chile, Colombia and Trinidad have prospects of increased
production based on natural gas.

Excluding Mexico, it is here assumed that net imports into the area will be no more
than 0.5 mn tons in 1985 and possibly lower by 1990.   Exports from Mexico are
forecast in the range 1.5-2 mn tons by 1985 and 2-3 mn by 1990.

Middle East.   Of non-Opec countries, Bahrain is a small exporter and Turkey the
only major importer.

In Bahrain the new Banagas plant has a capacity to produce 155,000 tons/year of
LPG and 125,000 tons/year of natural gasoline from 100 mn cu ft/day of associated
gas.   Approximately 40,000 tons of LPG from the Awali refinery is exported.   The
other major development in the area relates to Syria, where an LPG plant is under
construction in the northern oilfields.   The currently modest LPG consumption is
likely to rise to absorb most of the production from this plant as a substitute for
imported gas oil.   217

CHART 8.

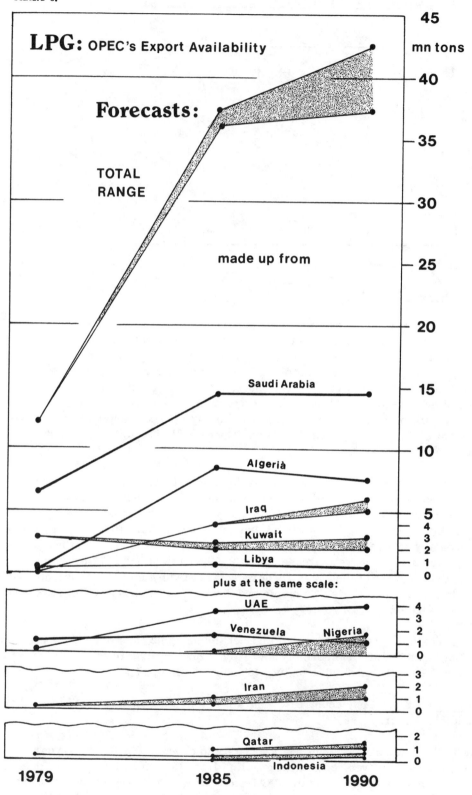

LPG: OPEC's Export Availability

45 mn tons

Forecasts:

TOTAL RANGE

made up from

Saudi Arabia

Algerià

Iraq

Kuwait

Libya

plus at the same scale:

UAE

Venezuela   Nigeria

Iran

Qatar

Indonesia

1979   1985   1990

Imports by Turkey are forecast as rising to 0.4 mn tons by 1985 and 0.5 mn in 1990, offset by exports from Bahrain of 0.2 mn tons.

Far East. The Singapore refineries are the major source of the modest amounts currently imported into other countries of the area. With the development of natural gas resources in Thailand and Bangladesh (in Malaysia and Brunei the LPG will not be stripped out of the exported LNG), the area could possibly increase its net exports, but in fact local consumption is more likely to rise to absorb the enhanced availability.

Oceania. Last year Australia's exports of LPG declined to 1.5 mn tons. Government fiscal incentives for the use of LPG as a motor fuel seem likely to reduce exports further. The 630,000 tons/year that it is proposed to strip out of North-West Shelf gas, as part of the project for LNG export, will not be available until the latter part of the 1980s. Much of it will be used locally. The South Australian government is still investigating a proposal to evacuate oil and gas liquids from the Cooper Basin to a fractionation plant at Stony Point, which could yield around 0.7 mn tons/year of LPG by 1984. Exports are estimated as remaining at 1.5 mn tons in 1985 and 1990.

New Zealand is planning to use its natural gas resources locally, to save on oil imports.

Africa. Both Egypt and Tunisia, currently the major importers, have projects under development and planned for LPG production based on associated gas. They will become self sufficient. The remaining countries of the continent are very modest consumers. In the case of East Africa, the combination of nearness to the Middle East, the lack of natural gas and the nature of the economies would seem to suggest that there is a potentially sizeable market waiting to be developed. It may, however, prove difficult for imported LPG to compete with what is invariably government-subsidised kerosine. Net imports in the area as a whole are forecast as falling to 0.3 mn tons by 1985 before recovering to 0.4 mn in 1990.

Summary. The table below sums up the above in respect of the future situation in non-Opec countries outside the communist bloc.

Rest of World (Excluding Opec and Communist Bloc):
LPG Export Availability and Net Import Requirements
(mn tons)

Export availability	1985	1990
Australia	1.5	1.5
Bahrain	0.2	0.2
Mexico	1.5-2.0	2.0-3.0
Total	3.2-3.7	3.7-4.7
Net imports required		
Central/South America	0.5	0.4
Middle East	0.4	0.5
Far East/Oceania [a]	-	-
Africa	0.3	0.4
Total	1.2	1.3

a Other than Australia.

CHAPTER 5

The Supply/Demand Equation

Introduction

The overall demand/supply situation in 1985 and 1990, as set out at the end of the preceding chapters, is summarised in the following table.

LPG Supply/Demand Balance in Non-Communist World
(mn tons)

Supply	1985	1990
Opec	36.1-37.4	37.3-42.5
Other	3.2- 3.7	3.7- 4.7
Total	39.3-41.1	41.0-47.2
Demand		
Western Europe	2.0- 3.0	2.7- 5.5
USA	4.8- 9.8	9.5-16.1
Japan	16.0-19.0	22.0-23.0
Other	1.2	1.3
Total	24.0-33.0	35.5-45.9
Surplus (deficit)	6.3-17.1	11.7- (4.9)

Surplus in mid 1980s –

A situation of LPG surplus in the mid 1980s has been widely forecast. As the table shows, even on the basis of the highs of demand in importing countries, the overhang in 1985 would amount to 6-8 mn tons. That is not a significant volume in the context of an oil demand in the non-communist world of perhaps 2,500 mn tons by then, but it implies that some LPG plants will not operate at full capacity[1], since the assumption underlying the high demand postulates is that price would fall to a level permitting them to be attained. Such a fall in price from present Opec postings may be achieved in one of two ways: a more modest escalation in LPG prices than in crude oil prices over the coming years[2], or a requirement that purchasers of crude oil lift cargoes of LPG. The effect would be the same in both cases since the latter is equivalent to adding a premium to the oil price.

---

1 The recently released Opec Annual Report for 1979 puts LPG availability in 1985 at 30.6-32.6 mn tons, the lower figures being essentially due to an estimate for Algeria of only 3.5 mn tons. In that event the surplus would disappear.
2 During 1980 average LPG prices moved down by $10/ton while crude oil prices rose.

CHART 9.

LPG: Supply and Demand

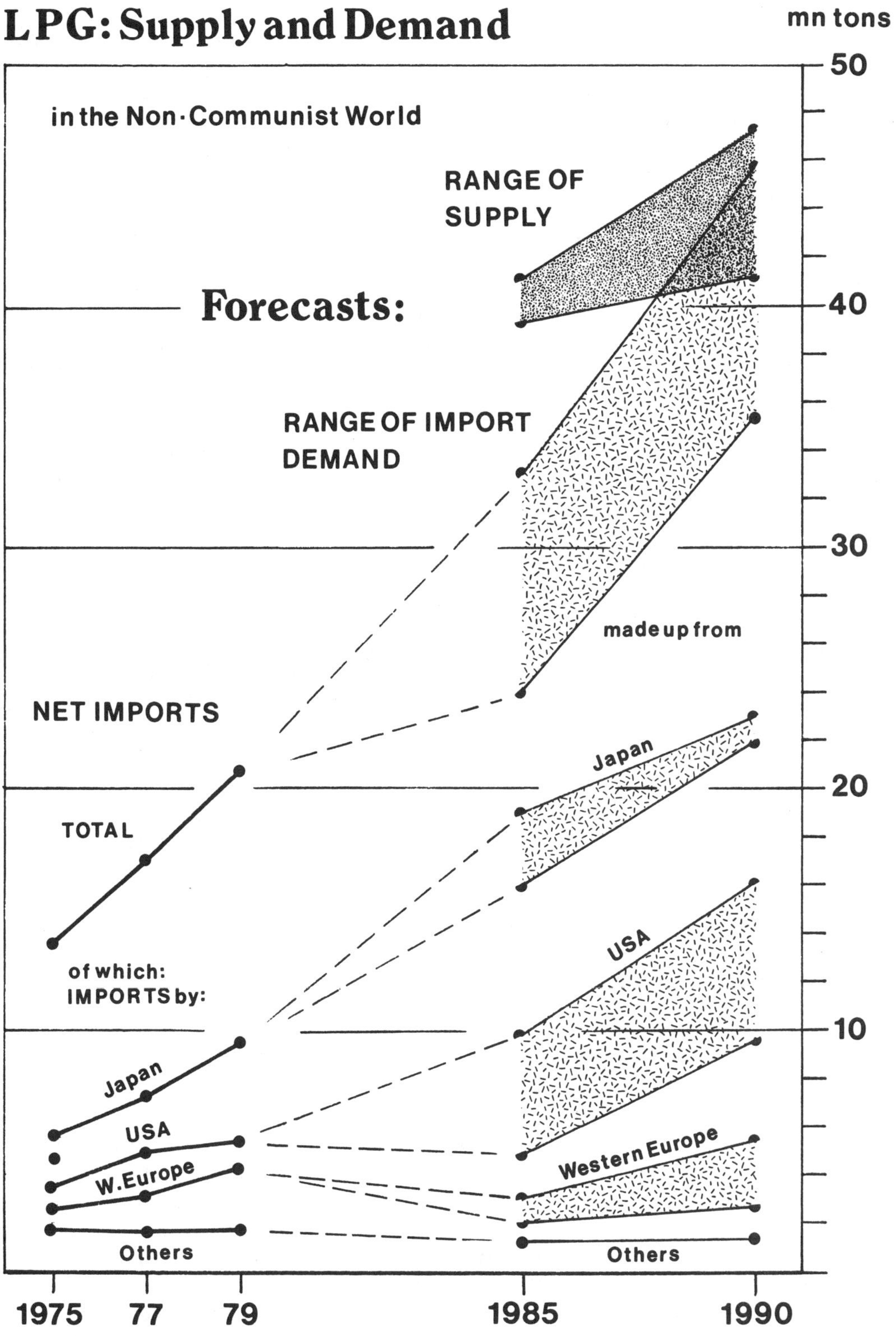

mn tons

in the Non·Communist World

RANGE OF SUPPLY

Forecasts:

RANGE OF IMPORT DEMAND

made up from

NET IMPORTS

TOTAL

of which:
IMPORTS by:

Japan

USA

W.Europe

Others

Japan

USA

Western Europe

Others

50

40

30

20

10

1975 77 79 1985 1990

- and possible deficit in 1990

The possible range of situations in 1990 is of more interest since it is the perception of what may happen then that will determine whether in the interim investment will be made, on the demand side, to substitute LPG for other fuels and feedstocks and, on the supply side, to increase production further. As regards the latter it may, for example, be postulated that if an overhang in supply does emerge in the mid 1980s, exports in 1990 are likely to be nearer the lower than the upper end of the indicated range. On the other hand - but for the same reason - the demand for LPG might be stimulated to the high of the range, in which case there would be a deficit, though not one of any significance. In this situation the US market is the key, and it is in the field of petrochemicals that the greatest scope lies for achieving a balance. Thus, for example, it is improbable that petrochemical producers in the USA will revert on any large scale to building plant to process LPG - as opposed, that is, to incorporating flexibility in existing and planned plants so as to vary feedstocks according to relative prices. Similarly in Europe and Japan.

The adjustment is not, however, as simple as this statement would seem to suggest. The use of LPG, whether as fuel or feedstock, requires special reception and storage facilities, and it is doubtful whether the necessary investment in installations will be made without some long term price commitment on the suppliers' side. Insofar as the petrochemical industry is concerned, there would be an advantage to the suppliers in making a commitment, since feedstock demand is less seasonal and erratic than fuel demand, and would thus represent a steady long term basis for sales. (On the other hand, those Opec members who are themselves envisaging large scale exports of petrochemicals may not find it attractive to benefit potential competitors in this way.) Experience in the mid 1980s might well incline producers to commit themselves in this way. If they do not, the situation in regard to LPG will continue to be similar to that currently facing the coal industry: there, too, it is generally agreed that consumption must increase to supplement static oil output, but hesitancy persists about making the necessary investment in terminals, storage and transport, as a result of both political and economic uncertainties.

Price relationships

In a situation made complex by the substitutability of fuels and feedstocks from different raw materials, their consequent price interdependence, and uncertainty about how the supply/demand for individual co-products produced from crude oil will balance themselves in the future, there are certain broad but fundamental relationships that have to be observed insofar as the pricing of LPG is concerned - and they apply whatever the price of crude oil may be.

1. In certain functions LPG can command a premium over the price of competitive fuels: certain industrial and agricultural processes where a clean fuel is required and for which natural gas is not feasible or available; remote or sparsely inhabited rural areas, mobile homes etc, to which it is not economic or feasible to pipe natural gas. Even in the second of these categories there are limitations on the premium that consumers will pay for ease of use over coal, kerosine or gas oil, while no premium would be acceptable over electricity. The point, however, about these industrial, agricultural and residential/recreational uses is that they have, in the past, been the market into which LPG has first flowed. In the

industrialised countries this premium market is now growing very slowly, and in some countries it is growing not at all as natural gas networks and electricity grids spread. Spain, for example, has been one of the fastest growing markets in Europe for LPG, essentially due to consumption in the residential sector; but that consumption is now static. Ostensibly a market exists in the developing countries. but there LPG has to contend with poverty, with a considerable fear of handling it (which the authors of this report found, in Nigeria for example, to be widespread, and not just confined to rural householders), and invariably government-subsidised kerosine.

The volume of LPG that is becoming available in the 1980s, though small in terms of oil, is nevertheless too great to be absorbed in these premium markets at a premium price. In other words, it cannot be this market that is used to determine the price of LPG in the future.

2. In other fuel uses (ignoring for the moment the special case of motor vehicles) LPG competes chiefly with natural gas and gas oil. There is currently considerable dispute in regard to the "just" relationship between the prices of natural gas and oil. It has, for example, been argued by Algeria that natural gas should be priced at its point of origin at calorific parity with crude oil. Not only does this thesis ignore the higher cost of delivering natural gas, whether by pipeline or as LNG, but it also ignores the fact that natural gas does not, at its point of use, compete with crude oil but, progressively as sales move down-market, with gas oil, low sulphur fuel oil and high sulphur fuel oil – how far down-market depends on the supply that has to be disposed of. At best, natural gas cannot be priced above gas oil equivalent in calorific terms at the point of use – and the same is true of LPG. In other words, the price of LPG netted back to its point of origin, whether local or overseas, must be lower than that of gas oil to an extent necessary to offset the higher costs of storing, transporting and distributing it. This equation clearly presents difficulties since it implies a range of netback values depending on the length of the supply/delivery trajectory. However, this geographical determinant is really no different in kind from the end-use determinant: if in a concentrated geographical area LPG is to expand its sales beyond the gas oil market, it has to be priced at a level that will permit encroachment on the fuel oil market – or at least priced at a level that will enable the distributor to average out the margins from his various offtakers. LPG, like natural gas and a number of other things in the world, can claim "nobility" only to the extent of its rarity.

3. Petrochemical producers do not buy calories but molecules. Very broadly speaking, therefore, they are not prepared to pay more for a ton of LPG than they would pay for a ton of gas oil, after allowing for differences in storage costs. Since there are approximately 10,650 kcal/kg in gas oil and 11,800 kcal/kg in LPG, this weight equivalence implies that a petrochemical producer will expect to pay less for a calorie of LPG than for a calorie of gas oil – approximately 10 per cent less. In other words, assuming that terminalling and storage costs are the same in both cases, the feedstock market does not in fact represent a premium market for LPG vis-à-vis LPG as a fuel competitor to gas oil. It can only become a premium market if different geographical situations alter the assumption of distribution etc costs being the same in both cases.

4. There is, finally, the special case of LPG as a motor fuel. To all appearances this represents a premium market since the competitor is gasoline, the highest priced of all oil fractions - the one that, since it has had no competitor, has always borne more than its fair share of refining costs. In fact LPG is never likely to disturb that situation: there is not enough of it. That in itself represents both the weakness and the strength of LPG: it has largely been ignored in the transport system, from vehicle manufacturers to service station operators, but it has also been ignored by governments on the prowl for new sources of revenue. The latter situation would probably not endure if LPG supplies were of such a volume as to affect the receipts from the gasoline tax. It is this circumstance which restrains car owners, and hence vehicle manufacturers.

Under certain fiscal conditions, as discussed in the chapter dealing with Western Europe, it is undoubted that motor fuel does represent a more attractive outlet for LPG than do most other uses. Even so, without yet further tax concessions than are usually available at present, the value of LPG in this market is still only around energy parity with gas oil - which is not to say that such concessions will not be granted, as the pollution-overwhelmed Japanese did in respect of the Tokyo taxi-fleet.

Appendix

APPENDIX

Table 1. Western Europe: LPG Consumption, Totals by Country
('000 tons)

	1967	1973	1975	1977	1978	1979
Austria	66	261	109	144	(138)	(131)
Belgium/						
Luxemburg	468	562	551	545	573	628
Denmark	180	303	183	201	208	230
Finland	(48)	90	87	95	119	128
France	1.737	2.696	2.589	2.703	2.892	3.084
West Germany	1.674	3.052	2.622	2.720	2,768	3,124
Greece	85	(125)	146	159	(165)	(165)
Ireland	41	92	104	128	(133)	(154)
Italy	1.360	2.051	2.118	2.216	2,245	2,241
Netherlands	259	469	472	781	1,162	1.868
Norway	18	28	28	36	582	(583)
Portugal	167	361	392	452	455	463
Spain	892	1,984	1,993	2,226	(2,239)	(2,388)
Sweden	(111)	190	187	170	191	196
Switzerland	27	83	75	70	(79)	(86)
UK	1,277	1,593	1,274	1,320	1,318	1.446
Total	8,410	13,940	12,931	13,966	15,267	16,916

( ) Partly estimated.

Table 2. Western Europe: LPG Consumption, Totals by End-Use
('000 tons)

	1967	1973	1975	1977	1978	1979
Gas manufacture	1,504	995	640	538	521	493
Road transport	313	659	824	1,196	1,296	1.375
Chemical/						
petrochemical	1,000	2,161	1,624	1,835	2,542	3.388
Other industry	1,310	3,335	3,118	3,522	3,751	4,048
Domestic/						
commercial	4,173	6,634	6,474	6,546	6,882	7,252
Agriculture & other	110	156	251	329	275	360
Total	8,410	13,940	12,931	13,966	15,267	16,916

Sources: OECD 1967-77. For end-use breakdown 1978-79: Finnish Petroleum
Federation, Central Statistical Bureau of the Netherlands, Norwegian Ministry of
Petroleum and Energy, Danish Petroleum Industry Association, Unione Petrolifera.
Fédération Pétrolière Belge, Svenska Petroleum Institute, Arbeitsgemeinschaft
Erdol-gewinnung und-Verarbeitung, Comité Professionel du Pétrole, Digest of UK
Energy Statistics. For other countries OECD totals have been used and end-use
breakdown estimated.

Table 3. LPG Consumption by Country and End-Use
('000 tons)

	1967	1973	1975	1977	1978	1979
A. Gas manufacture						
Austria	-	36	27	29	(29)	(20)
Belgium/						
Luxemburg	45	1	4	4	4	2
Denmark	20	30	18	17	16	34
Finland	-	-	11	11	11	11
France	200	361	292	238	217	188
West Germany	286	242	198	158	162	161
Greece	-	-	-	-	-	-
Ireland	-	-	-	4	(4)	(4)
Italy	20	5	5	10	-	-
Netherlands	10	1	-	-	-	-
Norway	-	-	-	-	-	-
Portugal	-	-	-	-	-	-
Spain	-	-	-	8	(8)	(8)
Sweden	(15)	25	22	25	33	26
Switzerland	5	32	10	8	(9)	(7)
UK	903	262	53	26	28	32
Total	1,504	995	640	538	521	493
B. Road transport						
Austria	-	2	-	14	-	-
Belgium/						
Luxemburg	33	18	32	39	36	32
Denmark	21	50	32	34	40	42
Finland	-	-	-	-	11	11
France	-	-	-	-	-	-
West Germany	15	3	3	-	-	-
Greece	-	-	-	-	-	-
Ireland	-	-	-	3	(3)	(3)
Italy	195	480	570	744	750	720
Netherlands	49	96	164	269	355	446
Norway	-	-	-	-	-	-
Portugal	-	-	-	-	-	-
Spain	-	(10)	22	93	(100)	(120)
Sweden	-	-	1	-	1	1
Switzerland	-	-	-	-	-	-
UK	-	-	-	-	-	-
Total	313	659	824	1,196	1,296	1,375

(continued)

Table 3. LPG Consumption by Country and End-Use (continued)

('000 tons)

	1967	1973	1975	1977	1978	1979
C. Chemical/petro-chemical industry						
Austria	16	154	4	-	-	-
Belgium/Luxemburg	25	20	17	13	7	27
Denmark	-	-	-	-	-	-
Finland	-	4	5	-	4	5
France	-	-	-	-	-	-
West Germany	817	1,628	1,313	1,381	1,299	1,452
Greece	-	-	-	-	-	-
Ireland	-	-	-	-	-	-
Italy	91	199	140	161	212	144
Netherlands	32	139	104	236	403	1,114
Norway	-	5	18	23	550	550
Portugal	-	1	2	2	-	-
Spain	-	-	3	-	-	-
Sweden	-	-	-	-	-	-
Switzerland	-	-	-	-	-	-
UK	19	11	18	19	67	96
Total	1,000	2,161	1,624	1,835	2,542	3,388
D. Other industry						
Austria	5	34	47	39	(41)	(32)
Belgium/Luxemburg	149	88	68	65	65	71
Denmark	26	130	53	77	104	100
Finland	(18)	66	31	50	68	76
France[a]	258	335	463	517	600	631
West Germany	250	697	(550)	500	532	530
Greece	8	(25)	28	32	(35)	(35)
Ireland	15	20	25	35	(35)	(40)
Italy	139	322	360	295	350	377
Netherlands	33	62	45	55	18	22
Norway	-	16	9	13	32	33
Portugal	11	70	78	86	92	(94)
Spain	-	(50)	66	453	(531)	(660)
Sweden	(96)	150	148	130	144	156
Switzerland	12	33	35	30	(35)	(39)
UK	290	1,237	1,112	1,145	1,069	1,152
Total	1,310	3,335	3,118	3,522	3,751	4,048

Table 3. LPG Consumption by Country and End-Use   (continued)

('000 tons)

	1967	1973	1975	1977	1978	1979
E. Domestic/ commercial						
Austria	45	35	31	62	(68)	(79)
Belgium/Luxemburg	213	419	420	410	447	479
Denmark	95	73	75	65	39	45
Finland	(20)	20	40	34	15	14
France	1,279	2,000	1,740	1,824	1,905	2,048
West Germany	255	462	(539)	666	(758)	(961)
Greece	77	(100)	118	127	(130)	(130)
Ireland	26	72	79	86	(90)	(107)
Italy	914	1,020	998	966	893	954
Netherlands	110	113	95	97	372	247
Norway	18	3	-	-	-	-
Portugal	156	284	309	360	363	(369)
Spain	892	(1,917)	1,893	1,672	(1,600)	(1,600)
Sweden	-	15	16	15	13	13
Switzerland	8	18	30	32	(35)	(40)
UK	65	83	91	130	154	166
Total	4,173	6,634	6.474	6,546	6,882	7,252
F. Agriculture & other						
Austria	-	-	-	-	-	-
Belgium/Luxemburg	3	16	10	14	14	17
Denmark	18	20	6	8	9	9
Finland	(10)	-	-	-	11	11
France	-	-	94	124	170	217
West Germany	51	20	19	15	(17)	(20)
Greece	-	-	-	-	-	-
Ireland	-	-	-	-	-	-
Italy	1	25	45	40	40	46
Netherlands	25	58	64	124	14	39
Norway	-	4	1	-	-	-
Portugal	-	6	3	4	-	-
Spain	-	(7)	9	-	-	-
Sweden	-	-	-	-	-	-
Switzerland	2	-	-	-	-	-
UK	-	-	-	-	-	-
Total	110	156	251	329	275	359

a Including chemical industry.

( ) Partly estimated.

DATE DUE